TOURISM, PERFORMANCE, AND PLACE

New Directions in Tourism Analysis

Series Editor: Dimitri Ioannides, E-TOUR, Mid Sweden University, Sweden

Although tourism is becoming increasingly popular as both a taught subject and an area for empirical investigation, the theoretical underpinnings of many approaches have tended to be eclectic and somewhat underdeveloped. However, recent developments indicate that the field of tourism studies is beginning to develop in a more theoretically informed manner, but this has not yet been matched by current publications.

The aim of this series is to fill this gap with high quality monographs or edited collections that seek to develop tourism analysis at both theoretical and substantive levels using approaches which are broadly derived from allied social science disciplines such as Sociology, Social Anthropology, Human and Social Geography, and Cultural Studies. As tourism studies covers a wide range of activities and sub fields, certain areas such as Hospitality Management and Business, which are already well provided for, would be excluded. The series will therefore fill a gap in the current overall pattern of publication.

Suggested themes to be covered by the series, either singly or in combination, include – consumption; cultural change; development; gender; globalisation; political economy; social theory; sustainability.

Also in the series

Tourism Destination Development
Turns and Tactics
Edited by Arvid Viken and Brynhild Granås
ISBN 978-1-4724-1658-2

Planning for Ethnic Tourism
Li Yang and Geoffrey Wall
ISBN 978-0-7546-7384-2

The Dracula Dilemma
Tourism, Identity and the State in Romania
Duncan Light
ISBN 978-1-4094-4021-5

Emotion in Motion
Tourism, Affect and Transformation
Edited by David Picard and Mike Robinson
ISBN 978-1-4094-2133-7

Tourism, Performance, and Place
A Geographic Perspective

JILLIAN M. RICKLY-BOYD
Indiana University, USA

DANIEL C. KNUDSEN
Indiana University, USA

LISA C. BRAVERMAN
Indiana University, USA

&

MICHELLE M. METRO-ROLAND
Western Michigan University, USA

ASHGATE

Published by
Ashgate Publishing Limited
Wey Court East
Union Road
Farnham
Surrey, GU9 7PT
England

Ashgate Publishing Company
110 Cherry Street
Suite 3-1
Burlington, VT 05401-3818
USA

www.ashgate.com

British Library Cataloguing in Publication Data
A catalogue record for this book is available from the British Library

ISBN: 978-1-4094-3613-3 (HBK)
978-1-4094-3614-0 (EBK)
978-1-4724-0374-2 (EPUB)

The Library of Congress data has been applied for.

MIX
Paper from responsible sources
FSC® C013985
FSC www.fsc.org

Printed in the United Kingdom by Henry Ling Limited, at the Dorset Press, Dorchester, DT1 1HD

Contents

List of Figures

List of Tables

About the Authors

Lisa C. Braverman is a doctoral candidate in the Department of Communication and Culture at Indiana University. In addition to the ways in which rhetoric intersects with tourism studies, her research interests include mobilities studies, biopolitics and necropolitics, and the troubled relationships between language and violence.

Daniel C. Knudsen is H.H. Remak Professor of West European Studies, a Professor in and Chair of the Department of Geography at Indiana University and Adjunct Professor of Anthropology and Tourism Studies. He is a cultural geographer working in the fields of tourism, landscape studies, and food studies.

Michelle M. Metro-Roland is Director of Faculty & Global Program Development at the Haenicke Institute for Global Education at Western Michigan University. Her research is situated at the interstices of cultural and urban geography and explores questions of landscape, material culture, and national identity. Along with several articles and a co-edited book, she is most recently the author of a monograph entitled *Tourists, Signs in the City: The Semiotics of Culture in an Urban Landscape*. She has an MA in History from UC Berkeley and a Ph.D. from Indiana University.

Jillian M. Rickly-Boyd is a Visiting Assistant Professor in the Department of Geography at Indiana University. She is a cultural geographer working in the fields of geohumanities and mobilities. Her research interests weave together: tourism studies, environmental perceptions, identity politics and biopolitics, and performance theories. From these perspectives, her work has given particular attention to authenticity in tourism motivation and experience.

Acknowledgments

Jillian would like to thank the many friends, colleagues, and teachers who have helped to shape her understanding of tourism and inspired her to think more critically about the field. Among the most formative, in this regard, have been my fellow co-authors of this book, as well as Charles Greer, Edward Linenthal, Dennis Conway, Dallen Timothy, Cem Basman, April Sievert, Paul Kingsbury, Ning Wang, Yaniv Poria, and Dean MacCannell. For their friendship and lively conversation, no matter what the topic, I wish to express my gratitude to Rebecca Clouser, Elizabeth Vidon, Shanon Donnelly, Jim Hayes, Bradley Lane, Brian Johnson, Todd Lindley, and Ben Shultz. For their support in making past research projects possible, I would also like to acknowledge the staff at Spring Mill State Park, Indiana University's College of Arts and Sciences, and the rock climbers of The Red. Further, much appreciation is due to Nicholas Wise, who graciously shared his bounty of photographs, collected from his travels across all seven continents, to help provide visual illustration to some of the concepts in this book. And, finally, I am thankful for my family, who has always believed in me, offering love and endless encouragement. Indeed, my husband, Bryan Boyd, is due the greatest acknowledgement, as he has never let me give up on anything, most importantly, myself.

Daniel wishes to thank his wife Linda without whom traveling would be no fun. Zack and Will have also been incredible sources of strength, pride, support, and inspiration for me. While short on commentary, Ted has been a good sounding board for many of my most harebrained ideas. I also wish to thank Charles Greer, Frank Hansen, and Pia Tripsen for their many insights into landscape studies and Denmark. This book would not be possible without my co-authors who have taught me more than I could have imagined about tourism, and who I am fortunate to count as friends. I also wish to thank my current and former students Anne Soper, Amy Savener, Gloria Maleski, Nitasha Sharma, Elizabeth Vidon, and Altynai Yespembetova for their insights and questions about my work. I am especially indebted to Anne Soper for introducing me to tourism studies. Finally, I gratefully acknowledge the H.H. Remak Professorship, the College of Arts and Sciences, the Institute for European Studies and the Office of International Programs at Indiana University for their continued support of my research.

Lisa wishes to thank the many people who have helped rework her research into Chapter 6. In addition to the co-authors of this volume, Charles Greer, Michael Kaplan, John Louis Lucaites, Phaedra Pezzullo, Susan Seizer, Jon Simons, Robert Terrill, Daniel Brouwer, Amira De la Garza, and Emily Cram have been immensely helpful in both conceptualizing my research and imagining the chapter. Leah Schneier and Dana Kamara generously offered their home in Israel so additional fieldwork could be undertaken. Graduate students from Indiana University's Departments of Geography and Communication and Culture, as well as Arizona State University's School of Human Communication, provided the emotional and intellectual support necessary to complete this project. Though too numerous to name here, these graduate students have become part of my family—and they have nurtured this work right along with me. My heartfelt appreciation to Norman, Linda, Ken, Dominika, Alex, Jerry, and Don Braverman, as well as the many kind and forgiving family members not mentioned here, for their support. Scheran Cox, Rachel Westberg, Gino Giannini, Rebecca Clouser, Todd Lindley, Elise Pang, and Genevieve Reisman have been tremendously generous with their time and friendship. Enduring thanks go to Chase Clow for being my most honest sounding board, and for always reminding me to smile and laugh.

Michelle would like to thank colleagues and teachers who, over time, have played a role in shaping her thoughts on semiotics, landscape, and tourism: especially Charles Greer, Ken Foote, Paul Groth, Eric Sandweiss, Tamara Ratz, Anne Soper, Yamir Gonzalez-Velez, Walter Roberts, James Hayes, and Shanon Donnelly. I also want to acknowledge the support from my co-authors and, especially, their limitless patience. And lastly, I am grateful for my partner, Dini Metro-Roland, for his willingness to debate epistemology and to journey with me through landscapes all around, even if a little *vakon*.

We could not have completed this volume without tremendous support from Ashgate—and especially Katy Crossan. Lastly, sincere thanks go to Barbara Schanel for providing thorough and helpful editorial advice throughout the project. As with most projects, this book necessitated the support of a whole community of colleagues and friends, for which we are all endlessly grateful.

Preface

We have chosen to write this book together, and have taken what we see as an honest approach—one that acknowledges each of our contributions and simultaneously acknowledges that we have relied on each other in formulating our separate ideas as we pursue a comprehensive, landscape-based theory of tourism.

This book also has a history. More than a decade ago, a student new to the Indiana University Department of Geography's graduate program arranged an appointment to speak with me about participating on her PhD committee. She was interested in examining the cruise ship industry in the Caribbean. At first, I declined to participate. I knew absolutely nothing about tourism studies. Yet, I was intrigued. And she was persistent. And so, with Anne Soper's help, I began my journey toward becoming a tourism scholar. At the same time, I also asked two colleagues to join me. The first of these was Charles Greer, who specialized in landscape studies. The second was Frank Hansen, who I knew from previous work in Denmark. Both were crucial to a landscape approach to tourism—Charles providing the necessary background in landscape studies and Frank, along with my Danish friend Pia Tripsen, helping me to think about it in place in the heaths, farms, and cities of Denmark. I was fortunate during this time to also work with Altynai Yespembetova, who forced me to think about how my Western ideas might play out elsewhere.

The ideas I had then were empirically driven and often not terribly well-focused. They needed formalization and theorization. It was at that very moment that the remaining three authors of this book arrived on the scene. The first to contribute was Michelle M. Metro-Roland, who set about formalizing the tourist experience using Peircean semiotics. Her work in Peircean semiotics is now central to how we all think about the tourist experience. Her discoveries also proved essential to much of the remaining theoretical framework on tourism sites. It was only after she had articulated the tourist-landscape experience that I was able to better theorize the tourism site, drawing heavily on modern theories of ideology and aesthetics. Yet, my understanding was incomplete in that the connection between physical object (site) and meaning/action was quite fragile. It was here that the research of Lisa C. Braverman, which draws from the field of rhetoric, became fundamental. Having theorized the tourist experience and tourism site, the problem of connecting the two remained. It is here that the work of Jillian M. Rickly-Boyd on authenticity, alienation, ritual, and performance is crucial. Her research on the many ways that tourism places the body in motion and emotion, while foregrounding relations of the self to place, works at the intersections of experience and site. The result, we think, is a more fully-fledged, landscape-based theory of tourism that effectively stitches tourist experience and tourism site together.

Daniel C. Knudsen

Chapter 1
Introduction: Landscape Perspectives on Tourism Places

Tourism Theories

Tourism is not a discipline, but a field of study. As such, scholars approach tourism from diverse perspectives, honing in on particular relations. Despite the greater attention paid to tourism across the academy in recent decades, we argue an inherent problem remains. Tourism researchers tend to narrow their focus, concentrating on either the tourist experience or the dynamics of the tourism site. Indeed, it is rare when a study is able to maintain an investigation of both aspects within the same framework. Further, we suggest geographers and geographic perspectives are particularly attuned to consider a holistic analysis of tourism. Tourism is a place-based endeavor. It occurs in particular localities, which are also set within broader geographic scales and networks of which tourist flows are just one of the many relations weaving through and connecting places. As such, we suggest, tourism performances do not begin and end in places, but through tourism, place is performed. Tourism fosters relations to and across places; it transforms places, engaging bodies, imaginations, and ideologies, well before and long after travel has occurred.

Looking at the tourism studies literature, one finds a number of edited volumes that have been able to speak across the tourist experience and the tourism site by fostering conversations among the numerous authors' research agendas (see Rojek and Urry, 1997; Coleman and Crang, 2002; Cartier and Lew, 2005; Minca and Oakes, 2006; and Knudsen et al, 2008). Yet, few monographs can be found that hold the two areas of tourism studies, site and experience, together while also moving theoretical discussions forward. Dean MacCannell (1976; 1989; 1999) was among the first to attempt this with *The Tourist: A New Theory of the Leisure Class*, the impacts of which are immeasurable in tourism studies. Significantly, he was among a group of tourism scholars readjusting perspectives on tourists by situating himself among those he studied. As opposed to some of the earlier tourism theorists, namely Boorstin (1961), MacCannell attested to the fact that scholars, too, are tourists. We do not simply sit in our ivory towers, studying the masses touring predetermined destinations, but we are active in this social process. Indeed this focus on tourists rather than tourism is of import. Engaging the theories of Marx, Levi-Strauss, and Goffman, MacCannell's work is particularly focused on the social structures that motivate and inform touristic experience. While he offers insights to the social processes of tourism sites—the ways in which destinations become attractions, the semiotics of sightseeing, and the staging of authenticity—his foregrounding of the tourist moves across the tourism site/tourist experience divide by following the tourist from home through the practice of sightseeing. His more recent book, *The Ethics of Sightseeing* (2011), further challenges the relations of tourists to attractions by calling into question the ethics and morality of tourism motivation and the social obligations that encourage travel.

Using the language of production and consumption of tourism, John Urry's (1990; 2002) *The Tourist Gaze* also holds the tourist experience and the site in tandem under a single theoretical framework. Complementary to MacCannell's work, this text pays greater attention to the economics of the tourism industry, including resort development, labor market and working conditions, and notions of spatial fixity. Urry works from the idea that tourism is about extraordinary experiences, that it is about being away from our everyday lives and encountering the "other." Inspired by the work of Foucault, this text extends the concept of the gaze to tourism, illuminating the power relations of tourist/site interactions. Urry's contention is that tourism is a primarily visual enterprise in which the gaze, a discursive relationship between viewer and object, informs both the production and consumption of tourism experiences. As such, he identifies a number of gazes—romantic, collective, anthropological, environmental, and others. So while Urry is able to theorize across the tourist experience and the tourism site, the main criticism of his text has been that it is too visually oriented, and as a result, it does not attend to the agency of locals, the subjectivity of being a tourist, and embodied practices of tourism. As such, it is important to note that the latest edition of *The Tourist Gaze* (2011), co-authored with Jonas Larsen, attends to such criticisms and more fully embraces a performative understanding of tourist practices, particularly photography.

A geographically diverse text, Edward Bruner's (2005) *Culture on Tour: Ethnographies of Travel* takes the reader through more than 20 years of his research. He shares a similar research agenda to MacCannell: to conduct an ethnography of tourism and touristic practices. Yet, as an anthropologist engaging a constructivist approach, Bruner "analyzes tourist performances not as representations, metaphors, texts or simulacra of something located elsewhere, but as social practice to be studied in its own right" (p. 7). This text is concerned with cultural tourism, focusing on the tourism performances that local/tourist/practitioner interactions yield, and giving particular attention to encounters, contestation, and narrative construction. This text moves across the tourist experience/tourism site divide of tourism studies by situating itself at the point of encounter. It is richly place-based in that regard.

Complementing these foundational works with a geographic perspective to both the tourism site and the tourist experience, we argue, will better account for the ways in which tourism spaces become places. While space is produced through practices, place is space made meaningful; it is more than functional, it is emotive and affective. For tourism, this means attending to the ways tourism spaces are socially produced but also the ways that tourism places become personal, embodied, and contested. Drawing together prominent theoretical themes of tourism research—ritual, semiotics, ideology, and performance—this text aims to hold together the tourist experience and tourism site dynamics as set in place. Indeed, Baerenholdt, Haldrup, Larsen, and Urry (2004) make important strides towards a geographic perspective in their text, *Performing Tourist Places*. Baerenholdt et al. state, "tourism is a way of being in the world" (p. 2). We must move beyond reading tourism landscapes to interrogate the touristic practices that continually (re)produce place. Their text provides engaging critiques that weave together theory and empirics as they investigate tourism performances. Thus, the emphasis on the performance of tourism underpins the ways in which these various theoretical perspectives (ritual, semiotics, and ideology) can be brought together. In this regard, we are referring to "performance" most generally, as the enactment of contexts and situations, events and moments.

Yet, a geographic approach to considering tourism destinations as places also necessitates a landscape perspective. Landscape, as a concept, allows for investigations of the social and cultural processes that shape and are shaped by place. It offers both outsider and insider perspectives, and incorporates various scales of interaction from the body to institutional structures of power. Indeed, it is for these reasons that Cartier and Lew (2005) have developed the concept of "touristed landscapes" as a means to account for the potential of understanding place as both toured and lived in. Tourism destinations are comprised of landscapes produced by and for tourism (see Aitchison, Mcleod, and Shaw, 2000; and Cartier and Lew, 2005). From the materiality of landscape to the imaginative, landscape is the medium with which tourists interact as they perform place. Further, it is landscape that tourism practitioners use in the marketing of tourism places. We see this text as working from and weaving together the rigorous and enduring theories of tourism, with other theoretical notes on tourism as well as those from outside of the field, by continually attending to the performative enactments of tourism places. This text, thus, works on various scales, moving back and forth from broader theoretical constructs (ritual, semiotics, ideology, and performance) to more specific relations grounded in tourism places.

Tourism Geographies: A Landscape Perspective

Landscape is a central concept in geography. As a discipline that includes both human and physical geography, landscape is a medium that holds the two in concert. While landscape may be examined as material culture, or what Carl Sauer (1925) stated as evidence of human agency on the natural landscape, Peirce Lewis (1979) suggests landscape is also our unwitting autobiography, a repository of social values and ideals. Denis Cosgrove (1984), however, asserts landscape is much more; it is a way of seeing. Perspective is essential to the understanding of landscape—insider and outsider—as landscape is a medium through which societies represent themselves to the world. Landscapes share but also extend the meaning of "area" or "region" by way of unifying the human subject and the material world. But landscapes not only reveal unities, they hide tensions (see Cosgrove, 1984; and Wylie, 2007).

Landscape as a construction, as a composition of the world, is also an artistic medium that employs single-point perspective to convey social relations. In graphic arts, perspective is used to portray distance on a flat surface, so that figures that would be further away from the viewer are smaller and those in the foreground, closer to the viewer, are larger. Perspective implies control, order, and distance, thereby locating the subject outside of the landscape, which representations of landscape then reflect back to the viewer. Thus, the study of landscape, in geography and elsewhere, has maintained a visual bias, a disinterested and supposedly objective stance with formal rules of investigation, which value outsider perspectives over insider experience by breaking the perceived unity of landscape into constituent parts of analysis. Such approaches to landscape study underwrite the very ideological position that landscape proclaims (Cosgrove, 1984). A humanistic approach to landscape, however, contends this is merely a surface-level investigation, below which lies much deeper meanings and tensions. To begin to recover the geographic imagination that inspires landscape construction and interpretation, we should also consider landscape as a literary device. As a text, landscape is a palimpsest, a layered record of social contests composed by a series of authors whose intertextuality conveys disparate meanings given particular readers (Duncan and Duncan, 1988). Indeed, Ponte Vecchio, Florence, Italy (Figure 1.1) illustrates landscape as a palimpsest in which multiple authors over time are manifest in present performances of place.

Landscape geographers have long danced around the issue of where meaning comes from in the landscape. Early comparisons of landscapes to texts suggested they might be read seemingly without problem, in a similar way to how one might read the words on a laundry list (Lewis, 1979). This critique is a false one since reading is always a more complexly negotiated task, engaging both one's ability to literally read the configuration of letters and requiring a far more nuanced ordering of information so as to actually understand what has been read. Just because one might be familiar with the Greek alphabet such that s/he can recite a sentence in Greek does not ensure that actual understanding has occurred. Similar to learning versus comprehending a language and, then, understanding the culture from which it has developed, landscape reading is fine-tuned over time. Yet, it is something we all do, that we must do, to navigate our daily lives, let alone when we are in unfamiliar environs. Surprisingly few are self-aware of their abilities to read landscapes, as such reading happens intuitively in most instances. Indeed, in concluding an essay in which he offered axioms for reading the landscape, Lewis (1979, p. 26) noted, "It is remarkable how many intelligent perceptive people have never asked questions of the landscape, simply because nobody ever suggested they do it."

Figure 1.1 Ponte Vecchio, Florence, Italy: Landscape as Palimpsest

In developing a critical perspective, the landscape-as-text approach in geography has focused on reading ideology and power relations in the (re)production of landscape. However, it has taken only minor account of pressing literary theory debates about the role of the text, the author, and the reader. In particular, Formalism and New Criticism have made little dent, as has the reaction to those schools of thought, Reader Response Theory (Eagleton, 1996). Nevertheless, some geographers have given more thought to the ways in which literary theory, generally, can be applied. Engaging Saussure and Barthes towards a post-structuralist approach to landscape as text, Duncan and Duncan (1988) argue landscapes are composed of signs working referentially. Yet, what landscapes reference is not always obvious to the reader, thus putting into practice the infinite deferral of the signified. As such, they suggest a landscape-as-text approach denies the authority of the author, and should attend to the intertextuality of landscape by focusing as much attention on the silences and absences as to what is revealed. While we agree with the premise and the potential of examining landscapes as texts, we have found a Saussurian approach to be limiting. In Chapter 3 we lend support to a Peircean semiotic approach to reading landscapes that accounts for the materiality of landscape, as well as the cognitive/imaginary aspects. Further, the Peircean approach to tourism landscapes offers a more open framework for considering the impact of guidebooks, promotional devices, and interpretative materials on the dynamics of tourist meaning-making before, during, and after touring.

Landscapes produced by and for tourism, perhaps more than our everyday spaces, work within the concept of genre. As Tzvetan Todorov (1990, p. 10) in writing about discourse argues, "literary genres, indeed, are nothing but such choices among discursive possibilities, choices that a given society has made conventional. [...] But there is no reason to limit this notion of genre to literature alone; outside of literature the situation is no different." Thus, consider the differences among the landscapes of adventure tourism, resort tourism, and urban tourism. Genres are established by constraints, by parameters that inform (re)production, reception, and understanding. They are embedded in larger webs of critique and historical context. In the case of tourism landscapes, they too are texts set within particular genres by the very fact that these landscapes are imagined, constructed, shaped, and manipulated to meet particular expectations of the audience and to support particular interpretations. More importantly, genres are not set in a vacuum but change over time, and thus, the landscapes of tourism genres also succumb to social and ideological changes.

Considering genre and landscape, (re)production moves the conversation importantly towards the role of discourse(s). The landscape-as-text approach outlined thus far, however, emphasizes an outsider's reading of landscape. While there is room for an insider perspective, interpretations and critiques continue to be made from a distance. And, indeed, while this orientation may be appropriate considering tourism landscapes are (for the most part) other-directed, we suggest acknowledging the scaling of landscape and utilizing both outsider and insider perspectives to attend to multiple scales of agency in landscape change and experience. In this vein, Richard Schein (1997) proposes landscape be approached as discourse materialized. He states, landscape is "a reflection of the people who created it, but as discourse(s) materialized, it also is implicated in the lives of those people and others in ways not accessible through such a traditional description" (1997, p. 666). By engaging the work of Foucault, Bourdieu, and de Certeau, Schein is able to articulate the ways in which the landscape holds different and competing meanings simultaneously, as well as how the "capillaries of power" that run through and circulate within landscape must be accepted, challenged, and/or modified through everyday practice. Such a conceptual framework acknowledges that landscapes are always in the process of becoming, and in so doing, provides a vision of landscape change that highlights the agency of the individual. Schein (1997, p. 676) suggests landscape interpretation should be about situating interpreters in place and broader geographic contexts: "the act of interpretation [...] should be processual and reflexive, open to the challenge of new information and alternative interpretations." Thus, in the case of tourism landscapes, attention to the discourses materialized highlights the multiple actors involved in landscape (re)production—locals, tourists, and tourism practitioners—and how discourses are both set in local contexts and informed by broader, global scales of power (see also Aitchison, Mcleod, and Shaw, 2000). As such, a "landscape as discourse materialized" approach is particularly applicable to investigations of tourism site dynamics.

Considering these developments for tourism studies, however, we are confronted with the fact that, as tourists, we do not simply look at and read landscapes from afar (Figure 1.2). We enter into them, we stroll, we lounge, we dine and chat with others, we absorb the sounds, smells, and sensations that surround us. As such, we also need a perspective that approaches landscapes from the most grounded level; that speaks to the experience of being in landscapes. Ingold (1993) argues a "dwelling perspective," informed by the works of Durkheim, Heidegger, and Merleau-Ponty, does just that. Ingold rejects the idea of landscape as a cultural image, arguing that it necessitates the division of inner and outer worlds, and instead, asserts landscape is the "domain of our dwelling" (p. 154). That is, landscape is constituted through continual acts of dwelling, always becoming, so that the body and landscape are complementary terms: each implies the other. Foregrounding agency and embodiment, a dwelling perspective, therefore, moves inward, providing an insider perspective to landscape so that there is no clear separation of self from scene. Agency, in a Latorian sense, means the ability to change a set of conditions; it is a capacity to act. There may be human or non-human agents, but in either case, agency presupposes reciprocity, that one moves with landscape, not across or through it. For tourism studies, a dwelling perspective offers a means to explore landscape/body relations, those moments of being a tourist (see Pons, 2003), in which embodied practices are the means through which one enacts place and self.

Figure 1.2 Florence, Italy: Landscape Perspective, Tourists on the Ground

By surveying these various approaches to landscape, we do not wish to suggest one is better than the others. Rather, our aim here is to elucidate the limitations of each approach. Each holds particular potentialities for the study of tourism. Further, considering multiple perspectives allows for a more robust accounting of the processes that drive, and are driven by, the (re)production, circulation, interpretation, and experience of tourism landscapes (see also Aitchison, Mcleod, and Shaw, 2000; and Cartier and Lew, 2005). So, while a landscape-as-text approach is a worthwhile approach to the analysis of tourism staging, the use of landscape in tourism marketing may be more comprehensively understood by considering landscape as discourse materialized. Further, addressing the tourist experience of being in place may require one to utilize a dwelling perspective that attends more to the body in relation to landscape. Thus, in this text we do not adhere to one approach to the study of tourism landscapes, but engage a landscape perspective that is open to the contextuality of the objects of study as we work to move across and maintain connections of tourist experiences and tourism site dynamics.

Place

Throughout this book we approach tourism from a landscape perspective, grounding tourism as a place-based endeavor. It is worthwhile, then, to provide an introduction to place as a concept. In the 1970s, Yi-Fu Tuan reoriented the way geographers understand and study place. As a result of the quantitative revolution in the social sciences, including geography, place had come to be considered as a specific point on a map or a particular location on the earth. Tuan, however, shed light on the experiential qualities of place. Most simply, he stated, place is space made meaningful. This occurs through time and experience with spaces. As such, Tuan (1974; 1975; 1977) articulates the role of passive modes of experience—smell, taste, touch—and active modes—sight and hearing. If the active modes of experience are the means by which we explore the world around us, the results of the passive modes of experience are inaccessible to others; they remain personal in terms of sensation and memory. Further, Tuan suggested that we construct place on various scales, from our home spaces to neighborhoods to cities to regions to nations. This scalar construction of place necessitates greater abstraction in shared experience and place associations the further one gets from the scale of the home. Tuan observed that home spaces are the most intimate because we experience them individually on a regular basis; similarly, neighborhoods and cities can be known through bodily experience. However, regions and nations are too large for one person to know intimately, and as such, politics, education, and the arts are employed to construct shared places at these scales. Thus, Tuan's (1974) *Topophilia* argues that place-making is essential to the human experience. We make meaning from the spaces we encounter. Some of that meaning is entirely personal, but much is shared.

While Tuan's work was essential to shedding light on the experiential notions of place, we can problematize his schema when considering tourism places. Tuan's articulation of the scalar construction of place and the necessity of greater abstraction through the use of art, education, and politics for constructing shared experiences of place can be observed. Yet, in tourism one is quite far from home while the bodily experience of place is foregrounded. One's inhabitance of a place may be temporary; nevertheless a personal and intimate relationship may be forged.

By the 1980s, cultural geographers were becoming interested in re-infusing the materiality of place with the experiential. In this vein, Agnew (1987) stated that Tuan's "sense of place" was just one element of place, along with location and locale. While location is the physical position on the earth's surface, locale is the material setting for social relations. Place, Agnew argues, is concrete; it has form. Even imaginary places of stories and dreams are situated. But place takes on meaning from social relations; a "sense of place" is the emotional attachment that results from this relationship. Such a perspective on place thus turned attention to the sets of relations that constitute place, but for the most part, place was still seen as developing out of "empty" space and then static once constructed on a social scale. As a result, this approach to place has a number of shortcomings for the study of tourism, namely that it does not account for non-sedentary conceptualizations of place-making.

The "new" cultural geographers and radical geographers of the 1990s, however, began to reassess the basic assumptions to understanding place. In particular, Doreen Massey's (1991) "global sense of place" challenged Relph's (1976; 1981) fears of "placelessness." Whereas Relph had suggested that increased mobility and global connections would diminish personal relations to places, as well as lead to the loss of place uniqueness through globalization of cultural forms, Massey (1991) argued that place is comprised of networks and flows, sets of relations that intertwine across and connect places—"what gives a place its specificity is not some long internalised [sic] history but the fact that it is constructed out of a particular constellation of social relations, meeting and weaving together at a particular locus" (Massey, 1991, p. 28). Such a perspective is, therefore, more applicable to tourism geographies as destinations become *meeting* places, which can be seen in Figure 1.3. This photograph of the streets of San Salvador, El Salvador illustrates the global and local networks that interconnect places, from tourism encounters to businesses (street vendors to multinational corporations). Instead of thinking of places as bounded areas, Massey states, place includes "a consciousness of its links with the wider world, which integrates [...] the global and the local" (1991, p. 28).

Building from the work of Massey, and increasingly useful for the study of tourism, is Thrift's (1999) relational materialist approach to place, which further attends to the role of embodied practice, agency of the non-human, and representation as doings. Thus, Tuan's modes of experience are still crucial to understanding place, but a post-phenomenological approach is further interested

Figure 1.3 The Streets of San Salvador, El Salvador. Place as *Meeting* Place (by Nicholas Wise)

in the ways in which sensory modes are engaged in breaking down subject/object binaries and establishing a relationship to place (see Casey, 1993; 1997; and Malpas, 1999). Cartier and Lew (2005) argue that sensory modes are employed by tourism practitioners to entice tourists by conjuring potential experiences, thereby engaging the seductive possibilities of place.

Central to such a relational materialist approach to place, states Simonsen (2008, p. 15), are the notions of practice, process, and change—"places are never finished but always becoming." However, such an approach also reconsiders space, not as empty or abstract, but as produced through social practice (see Lefebvre, 1991). Place does not emerge out of "nowhere," but it is a personal experience of the networks of social relations that interweave through and across spaces, changing over time—forged, consolidated, eroded, and renewed (Baerenholdt and Granas, 2008). Places are enacted; they are points of encounter, mooring, and connection. So that even on the rigid itinerary of a bus tour, places become connected, linked to previous destinations in the course of travel through new landscapes, but the points of mooring offer intimate connections with locals as well as other tourists (Figure 1.4).

By tracing the developments in conceptualizing place, we wish to advocate for an opening up of place as points of connection and encounter. Such an understanding of place builds from traditional accounts of place as locational and static to better

Figure 1.4 Iceland Tour Bus: Tourism Places as Points of Encounter (by Nicholas Wise)

attend to the ways increasingly mobile societies know and experience place. The relationship of place and landscape is, thus, well summarized by Ingold (1993, p. 155):

> [A] place in the landscape is not 'cut out' from the whole, either on the plane of ideas or on that of material substance. Rather, each place embodies the whole at a particular nexus within it, and in this respect is different from every other. A place owes its character to the experiences it affords to those who spend time there—to the sights, sounds and indeed smells that constitute its specific ambience. And these, in turn, depend on the kinds of activities in which its inhabitants engage. It is from this relational context of people's engagement with the world [...] that each place draws its unique significance. Thus whereas with space, meanings are *attached to* the world, with the landscape they are *gathered from* it.

Working from a landscape perspective and considering tourism spaces as places, we argue, offers a means of integrating the tourist experience and the tourism site as a performance. A landscape perspective attends to the social relations that produce and are produced by the built environment, as well as a way of seeing both the insider and outsider perspectives. Conceptualizing tourism spaces as places, then, attends to the experiential elements, as well as the contextuality and connectivity of places. Engaging a geographic perspective on tourism places foregrounds the ways in which the politics and poetics of space are enacted.

Structure of the Book

This book is divided into two parts, each containing four chapters. Part One establishes the theoretical frameworks from which tourism can be approached—ritual, semiotics, ideology, and performance. The fact that each chapter focuses on a particular theoretical aspect of tourism is an effort to highlight both the ability of scholars to focus on a particular perspective but also to illuminate the limitations of such an approach. As we hope will become increasingly clear to the reader, each theoretical perspective—ritual, semiotics, and ideology—works relationally to the others in performative ways. That is, an investigation of tourism from one of the perspectives immediately draws the others into the conversation as tourism is a social phenomenon that is performed, that is put into practice, by tourists, locals, and tourism practitioners. As such, we have made efforts to clarify conversational threads that run through and across individual chapters.

Part Two grounds tourism performances in places as a means to illustrate the ways in which ritual, semiotics, and ideology are relational in performances of tourism. Each chapter foregrounds specific processes, but attends to performative enactments of place. Further, each chapter stems from a different author's research interests. This approach has proven both fruitful and limiting in the context of a book interested in tourism theory. Drawing on each of our contextual research interests offers the reader rich, grounded investigations of tourism in place. However, the use of our personal research has also resulted in a rather limited, predominantly Western perspective. While we recognize the shortcomings of case studies that do not attend to tourism performances outside of North America, Europe, and Israel, we are confident that the theoretical frameworks are relatively generalizable across tourism destinations.

Chapter 2 examines the ritual aspects of tourism as a precursor to the contemporary studies of tourism performances. As with the entirety of this volume, this chapter importantly draws on the work of MacCannell and extends his definitions of sightseeing and tourism. We emphasize that tourism does not end at the destination, but after one's return "home." Despite the many idiosyncrasies of individual journeys—delayed transportation, peculiar friendships formed in transit, or an especially outstanding or disappointing meal—tourist experiences often follow a widely generalizable, ritualistic pattern. Chapter 2 investigates the components of this pattern, as well as the theoretical import of ritual in tourism. As tourism moves from a distinctly modern to an increasingly fragmented, postmodern endeavor, we question to what degree the social functions of its rituals have changed. Yet, this chapter also works to ground rituals of tourism in places. Returning to religious studies, where many theories of ritual originated, we articulate the parallels of the ritual production of sacred space to the ritual production of tourism space by drawing attention to practices of consecration, defilement, hybridization, and exclusion.

Chapter 3 explores the intersection of the tourist and the site as an experience of place using Peircean semiotics. Though the embodied, conscious, and conscientious process of constructing and interpreting signs is not exclusive to tourism, the often unfamiliar surroundings of tourism destinations bring this process to the fore. Further, tourists regularly confront materials (promotional and interpretative devices) that work to shape and limit interpretations of place. Building upon the work of Metro-Roland (2011), this chapter highlights the distinctiveness of Peirce's semiotic. Peirce contends that signs implicate actions and habits in addition to cognitive associations. Furthermore, Peirce's work describes interpretation as an ever-evolving, never fully completed phenomenon. This particular theorization of semiotics is invaluable to the study of tourism, as it helps articulate the intersubjectivity of semiosis as well as the delicate interplays between the body and mind while one is on tour.

Chapter 4 examines ideology's implications for tourism. Reiterating the dialectical tensions between the universal and the particular we began to animate in Chapter 2, we explore how cross-cultural tourism may invite a diversity of interpretations and opportunities for tourist agency. Although some tourist sites emerge quickly and organically from visitors (for example, the location of a natural disaster that might inspire voluntourism), most are chosen by host cultures. Ideology and myth are employed to attract tourists to certain locations and attempt to reify (host) cultural identities so tourists might gaze at them. The reception of ideological signs is neither simple nor unilateral, however. Tourists possess individual histories and sign associations; this knowledge may clash with the ways in which host cultures wish themselves to be perceived. Issues of ideology in tourism raise questions about national identity, site selection, and the degree to which an individual can direct his or her tourist experience.

Chapter 5 provides an overview of ways performance theory has been and could be applied to tourism. First examining tourism through the metaphor of the stage, we consider the relationships between guides and members of host cultures (performers) and tourists (audiences), while noting that these categorizations are neither tidy nor mutually exclusive. Performance in tourism extends beyond staging, however, and we proceed to explore performance and tourist agency. Tourism is, in fact, an embodied performance of place that may be analyzed through lenses of non-representational theory and persuasion, among others. Tourists themselves enact place identities, and the blending of tourism studies and performance theory provides a single framework through which scholars might examine both the tourism site and tourist experience.

Chapter 6 analyzes the ways peculiarly embodied, performed persuasion can render the ideology of the tourist experience more difficult to refuse. Based upon fieldwork conducted on a Birthright Israel tour, this chapter analyzes movement and militarization as staged tactics that may be employed on a tour. Birthright Israel is a free, ten-day tour for Jewish young adults, meant to (re)connect them to their Jewish identities. Though unique in its aims, the enthymematic structure of this tour may be applied more expansively. In an enthymematic tour, tourists complete important communicative events that have been started by their leaders. Working through several intersections of ritual, semiotics, ideology, and performance, this chapter alludes to issues of pre-trip marketing, notions of belonging, and what can be defined as "home," all of which are expanded upon in subsequent chapters.

Chapter 7 considers tourism promotion and onsite brochures using ideas drawn from Chapters 3 and 4 to problematize tourism promotion and site/experience interaction. In so doing, we illustrate our points using an analysis of a Danish tourism promotion video. This is followed by a content and semiotic analysis of tourism literature collected at sites in Denmark. We argue that a variety of disconnects emerge in promotional materials. In the video, often unrecognizable signs are used to entice potential tourists. The site brochures we examine clumsily handle language barriers, reducing content in translation, shifting meaning, and providing insufficient context for chosen words.

Chapter 8 expands upon the points made in Chapter 2 regarding tourism as secular ritual, and explores how communities are formed in those rituals. We demonstrate how theoretical notions of ritual, liminality, and communitas are intertwined at various stages of touring. This chapter dwells on ideas of communitas as they played out in two distinct destinations, Venice and Florence, Italy. Venice and Florence represent two different iterations of communitas: a more frequently visited space in which fields of care have begun to develop, thus generating a feeling of community, and a locale of more spontaneous and fleeting connection and identification.

Chapter 9 pushes the definitional boundaries of tourism by calling into question the concept of home. Also engaging performance-based theories, this chapter investigates a form of lifestyle travel—lifestyle rock climbers. Driven by a passion for rock climbing, these travelers maintain hypermobile lifestyles; they typically utilize a van as a common abode, thus exemplifying the precondition of privilege. While continuously on the move, lifestyle climbers forge strong connections to place and community, and in so doing, continually and habitually produce home on the road. This chapter works at the border of tourism studies and mobilities studies to open up the concepts of home and place, as a means to better account for individual and collective performances of community, belonging, and dwelling.

In Chapter 10 we provide a summary and conclusion to this book. Additionally, we suggest ways in which further tourism theorizing might prove fruitful by pushing the boundaries of tourism outward to connect with other bodies of critical humanistic and social scientific theory.

PART I
Theoretical Frameworks

Introduction

This book is inspired by classic theoretical texts in tourism studies, those scholarly works that have developed theories which encompass touristic experience and tourism site dynamics. Dean MacCannell's *The Tourist*, a crucial text upon which we build in this book, offers an ethnographic examination of modernity that hones in on the act of sightseeing as exemplary of the era. Following the tourist from the alienation of everyday life through the ritual attitude of tourism, MacCannell also explicates the semiotics of attraction and the staging of authenticity. John Urry's *The Tourist Gaze* further interrogates the notion of tourism as extraordinary, compared to our ordinary everyday lives, thereby considering the production and consumption of tourism with greater attention to the economics of tourism as an industry. In addition to these two foundational texts of tourism theory, Nelson Graburn's (1983; 1989; 2001) work regarding the ritual aspects of tourism, Edward Bruner's (2005) rich investigations of the politics of tourism encounters, and Erik Cohen's (1973; 1979; 1992) theories of touristic motivation and experience have been essential to thinking more holistically about tourism. Building from existing theories of tourism, this book identifies four broad theoretical constructs—ritual, semiotics, ideology, and performance—that each hold the tourist experience and tourism site together while also necessitating others. In Part One, each conceptualization is the focus of a single chapter, but as will become apparent to the reader, a discussion of tourism requires a conversation among many theoretical approaches.

We begin, in the following chapter, with ritual as a way to excavate some of the roots of tourism theory. MacCannell (1976) identified sightseeing as a ritual of modernity, working in contrast to the theorization of Boorstin (1961), which stated that tourism is composed of pseudo-events in which tourists seek inauthenticity as a justification for inauthentic lives. As such, MacCannell was suggesting a convergence of the social roles of tourism and pilgrimage (Cohen, 1992) in that tourism serves a broader social function. The ritual attitude of tourists and the processes of sight sacralization offer a means of social integration, a way of alleviating the alienation of everyday life. Graburn (1983) further extended the scale of analysis to consider the entirety of the tour as a ritual. He also situated his theory of tourism as "secular ritual" more squarely in theories of ritual derived from religious studies, thereby identifying structures and phases of this ritual—

pre-trip and departure rituals, liminal period of travel, re-entry rites. Yet, taking a geographic perspective on the ritual aspects of tourism draws attention to tourism performances of places. As such, we bring MacCannell's ritual attitude of sightseeing into conversation with Graburn's secular ritual of tourism and theories of sacred space, which suggest that sacred space is ritualized space. From this perspective, a number of similarities can be observed in the processes that constitute tourism places and sacred spaces, namely processes of consecration, interpretation, hybridization, and purification. Particularly important in this regard is the understanding that rituals are performative, communicative acts. It is the performance of tourism rituals that (re)produce tourism spaces.

The emphasis on ritual, and ritualized spaces, as performative and communicative necessitates the inclusion of semiotics and ideology in a conversation of tourism theory as it calls into question for whom a space is performed. Semiotics is also deeply entrenched in the earliest theories of tourism. MacCannell's semiotics of attraction employs both Saussurian and Peircean semiotics to expound the relations of marker/sight/tourist in the construction of a tourism attraction. Since then, semiotics has been more fully fleshed out in relation to the politics of tourism and, indeed, its potential for tourism research is still being realized. Thus, we detail some of these moments in tourism studies in Chapter 3 while lending support to the semiotics established by Charles Sanders Peirce. Peircean semiotics proves particularly useful in maintaining a geographic perspective on tourism as it maps well on to theories of landscape-as-text. That is, while tourism involves the performance of places, tourists are not bound to and in the spaces of tourism; they bring with them preconceptions, expectations, and other collateral information from past experiences that are put to use in reading the landscapes of tourism destinations. While landscape reading is something we all do every day, in unfamiliar spaces the processes of semiosis become all too apparent as we attempt interpretation. Yet, semiosis is not an individual endeavor. Our collateral information is shaped by knowledge gained from other people and sources. This is particularly true of tourism in which guidebooks, promotional materials, and word of mouth accounts inform expectations. Further, on the ground, interpretive materials and intra- or cross-cultural observations refine interpretations and thereby influence one's actions.

Thus, such a discussion of semiotics and tourism eventually turns to notions of ideology, particularly in regards to the ways tourists' interpretations are influenced, shaped, and limited. We frame our discussion of ideology in Chapter 4, however, with an incredibly broad and overarching question for tourism studies, one that harkens back to the works of both MacCannell and Urry—why and how do some sites become tourism destinations and others do not? While a comprehensive answer to this question is hardly possible, raising it foregrounds more specific questions regarding the institutionalization of tourism destination selection and promotion, the mechanisms employed in marketing of tourism, and the role of tourist agency in delineating tourism spaces. Identity politics are central to tourism. In the case of tourism site selection and promotion, particular organizations are established in order to choose among innumerable potential sites to highlight those that speak to place-identity best, that is, those that support a particular construction of place-identity. The ideological implications of tourism site selection continue to the marketing of destinations. Working to construct place myths, tourism marketing uses aesthetics to foster sign associations as a means to entice the desire to visit and influence tourists' expectations of place. Once on the ground, tourists may be presented with interpretative materials intended to guide specific understandings and readings of place. Yet, tourists maintain agency even in the most staged of destinations. Tourists perform places, and it is in these performances that ideologies can be accepted, challenged, or rejected.

This discussion of the theoretical constructs of tourism reveals how ritual, semiotics, and ideology relate to one another, most importantly, in performative ways. The chapter on performance theories in tourism studies comes at the end of Part One so as to lay the groundwork for the other theoretical constructs, which have a deeper history in tourism studies, and so that an explicit discussion of performance theories can establish the current state of the field and future trajectories. Of course, some connections to foundational works must be delineated, and indeed, MacCannell's work is of import here. Goffman's theory of staging in social interaction is crucial to MacCannell's notion of staged authenticity—the various ways in which tourism encounters are framed—front regions (where hosts and guests meet) to back regions (where work is completed and thereby outside of tourist access). We revisit Goffman's work on framing and footing to problematize dramaturgical metaphors so as to better articulate the politics of tourist/site interaction, namely persuasion, by considering cases of guided tours and activist tourism. Further considering the performance of tourism from a geographic perspective turns attention to landscape and place. Recent work with non-representational theories suggests the embodied practices of tourists enact place. Going beyond the gaze to the body in the production of space, performance theories attend to the ways landscape continually unfolds with movement, always anew. Such perspectives also shift greater emphasis to tourist agency in the performance of representations, that is, on the (re)production and circulation of ideologies.

Performance theories in tourism studies build from previous understandings of ritual, integrating semiotics and ideology in increasingly complicated, yet productive, ways. Considering rituals as performative and communicative acts, semiosis as the interpretation and communication of signs, and ideology as specific power relations and political agendas, performance is essential to each and holds them together. Tourism places are performances of ritual, semiotics, and ideology that do not occur solely on the ground in tourism destinations. These performances begin well in advance and extend long after the tour itself, so that tourism places become parts of individual networks or constellations of places.

Chapter 2
Rituals of Tourism

Introduction

From metaphors of staging, to rhetorical critiques of tours, to embodied practices, tourism is a performative series of rituals. Research focused on tourism as performance is a more recent development (see Chapter 5); however, the roots of this trajectory run deep in tourism studies. Performances are communicative acts, central to which are rituals, gestures, habits, and other embodied practices. Such a combination of acts suggests why some of the earliest theories of tourism concerned ritual aspects, as rituals are socially performative—and therefore communicative—devices. Furthermore, notions of ritual signify the ways in which tourism practices persist through time, almost but not completely replicable in a multitude of circumstances. MacCannell (1976) in particular noted that ritual aspects of sightseeing perform a social function as they offer tourists the potential for social integration. Traveling to "must see" attractions, the individual participates in a collective ritual distinctive of late modernity. Graburn (1983) then extended MacCannell's analysis to the rites and rituals of tourism generally, taking note of pre-trip rituals of travel preparation, the liminality of the trip, culminating in the return home and rites of re-entry to the everyday. Over time these ideas have been further expanded. As a result, tourism scholars have undertaken more nuanced examinations of tourism performances. This chapter examines the predecessors of performative perspectives in tourism studies, namely ritual, so as to establish the foundation on which current research in the field builds. In so doing, this chapter also works to elaborate on analyses of ritual by incorporating theories overlooked or underexplored in earlier scholarship.

Ritual

Theories of ritual derive primarily from religious studies. Rituals are first and foremost performed; they are "a particular type of embodied, spatial practice" (Chidester and Linenthal, 1995, p. 9). Additionally and more specifically, rituals are noted for serving a double social function—integrating individuals into a social structure and stitching together social divisions (see van Gennep, 1960; Turner, 1969; and Smith, 1987). Theoretically, the core characteristics of ritual include repetition and the ability to transmit symbolic codes within a cultural group. MacCannell (1992) problematizes both of these aspects, however, and in so doing changes the ritual perspective, making it more applicable to postmodernity broadly and tourism studies specifically.

First, MacCannell (1992, p. 257) argues that while "ritual performances are framed as pre-set," they are in fact more accurately occasions of failure to uphold a standard form. According to Turner (1982, p. 79), "rules 'frame' the ritual process, but the ritual process transcends its frame." In fact, MacCannell suggests ritual performances are better understood as a privileged space in which symbols may be modified. Thus, while rituals are thought to be repetitive, they are actually never truly repeated. This lack of perfect replication can be seen, among other places, in rituals that accompany national celebrations as well as personal preparations for travel. These rituals are circumstantial, contextual, and highly adaptable. The pre-departure rituals analyzed in Chapter 8, for example, are not repeated exactly each time a particular couple decides to travel internationally. Each trip, even if to the same destination, often requires different purchases and mental preparations. In the case of the trip to Florence in Chapter 8, while not this couple's first time to Europe, lessons learned in previous travels inspired the purchase of new luggage for this particular tour. More broadly, the rituals of tourism involve different actors, dangers, and personal concerns.

Second, more than adaptation and change that occur over time, MacCannell argues rituals *must* change, as pure repetition actually degrades the ritual. To return to Turner (1982, p. 79), to perform is "to bring something about," but in carrying out a performative act, "something new may be generated" and as such "[t]he performance transforms itself." This suggests that, most generally, rituals are communicative performances that are informed by social structures and aim to clarify intentions. Imperfect replications of rituals always undergird social change, yet one might wonder how this framework might be useful to tourism studies. MacCannell (1992, p. 271) argues that a more open reading of ritual illuminates the everyday manifestations of it, and moreover, its politics—every encounter "can and does become an occasion for transgressions of others' subjectivity." Further, this perspective foregrounds the creative potential of ritual performance rather than its deterministic nature, as previously understood. Ritual, somewhat like the enactments of ideology described in Chapters 4 and 6, presents unexpected opportunities for agency. Insofar as rituals evolve and change, simultaneously enforcing norms and transgressing them, they are generative rather than repetitive. For example, as of this writing, the U.S. Transportation Security Administration (TSA) has recently lifted the ban on select electronic devices being used during airplane takeoff and landing. In-flight rituals changed almost immediately, and generated significant publicity while doing so.

From these critiques, MacCannell (1992) also suggests that considering ritual in this generative light better integrates it into a semiotics of tourism. Using Peircean semiotics, he argues that ritual, like a sign, "address[es] somebody, that is creates in the mind of that person an equivalent sign" (1992, p. 275). He takes this idea a step further, however, and suggests we ask, in regards to ritual performance, "Who is speaking, and to whom?" (p. 275). This is a question we interrogate in the next two chapters. As evidenced by the logical fallacy of the equivocal sign, signifiers and signifieds do not necessarily remain stable throughout history. Signs change at such a rapid rate, interpreting them as stable across months, years, and decades often results in analytical error. Questions of signification are taken up more explicitly in the next chapter.

Sightseeing as Modern Ritual

The concept of alienation is at the heart of the earliest theories of tourism. It is offered as a motivating force behind touristic seeking of authenticity as well as efforts at social integration through the ritual performance of tourism. The concept of alienation has roots in both Marxism and Existentialism, two primary philosophical influences in tourism theory (see Rickly-Boyd, 2013). Both developed from Hegel's alienation—the separation of consciousness from reality. In Marxism, alienation is the result of separation of the self from the products of one's labor, from one's fellow workers, and the State. Among Existentialists, alienation is taken to mean a denial of one's self, be that through stifling attempts at social conformity or the overwhelming existential angst that accompanies attempts to express one's authentic self. Xue, Manuel-Navarrette and Buzinde further delineate a consumer alienation characterized by a state in which commodity fetishism is aided by advertising and mass media so that individuals "actively engage and integrate themselves into a hegemony of consumption" (2014, p. 190).

MacCannell's (1976; 1999) germinal text is among the earliest discussions of alienation with regard to tourism motivation. He notes that there appears to be a kind of "miracle of consensus" to what constitutes the most important tourism attractions. It is from this observation that he excavates the institutional mechanisms—processes of sacralization and ritual attitude—which act in tandem to (re)produce the objects of tourists' sightseeing. This ritual attitude surfaces in modern society as a sense of duty. The definition of ritual used in this context is derived from Goffman "a perfunctory, conventionalized act through which an individual portrays his [sic] respect and regard for some object of ultimate value to its stand-in" (1971, p. 62 cited in MacCannell, 1999, p. 42). Sightseeing, MacCannell argues, serves as an example of ritual attitude *par excellence*, as the ritual act originates with the initiation of travel and culminates with the arrival at the sight of interest. The performance of this ritual offers individual tourists the potential of social integration (1999, pp. 43–46). "Sightseeing is a kind of collective striving for a transcendence of the modern totality, a way of attempting to overcome the discontinuity of modernity, of incorporating its fragments into unified experience" (1999, p. 13). Indeed, tourism is a striving. While tourists attempt to construct totalities through collective ritual, the persistent practices of touring actually celebrate differentiation by setting aside attractions (see Chapter 3).

But how do tourists come to know which sights are "must see"? This aspect of sightseeing is theorized as a process of sacralization, which in the modern era comes about via extensive institutional support with clear structural elements. As noted in Chapter 4, ideological and hegemonic apparatuses undergird many selections of tourism sites. MacCannell identifies five stages of sacralization: naming, framing and elevation, enshrinement, mechanical reproduction, and social reproduction. Through this process sights are marked, reproduced, and circulated as worthy of special attention (see Chapter 5 for further discussion and an example of sacralization). While seemingly arbitrary, MacCannell (1999, p. 45) argues these representations are not random collections, but tend toward universality. The process of sight sacralization is one of the ways the ritual attitude surfaced under modernity; social integration had not decayed, as Goffman and Levi-Strauss contend, but happened at the macrostructural level (1999, p. 46). Though many tourism rituals embrace cultural specificity, their proclivities to wide appeal and interpretation ensure that rituals are not confined to particularity.

Theories of the ritual of sightseeing proposed by MacCannell explicitly focus on the modern era. Yet, the postmodern tourist (or post-tourist) is an entirely different creature, illustrating a greater attunement to self-reflexivity while playing with dualisms of modernity, such as subject/object dichotomies (Oakes, 2006). Recent work, however, offers evidence that the underlying mechanisms of MacCannell's theories, notably alienation and ritual, hold strong for post-tourists, despite different tourism performances and goals for travel. In recent research on the circle of representation in tourism imagery and touring practices, Jenkins (2003) contends we can see the ritual attitude of tourism at work. While Jenkins uses Urry's (1990) notion of a hermeneutic circle to analyze her findings, the evidence presented also moves MacCannell's theories of ritual attitude and sacralization into *post*modern tourism. A circle of representation is a process in which images presented to potential tourists through destination marketing imagery are the same sights visited, and images (re)captured in touristic photography practices (Figure 2.1). Focusing on backpackers, the ultimate post-tourists who make explicit attempts to avoid anything "touristy," Jenkins finds these tourists similarly participate in circuits of reproduction. The Australian tourism brochures she examines are produced in several forms so as to attract various types of tourists, with those aimed at backpackers appearing more casual in layout and composition of images. For example, the covers of brochures directed towards backpackers present a collage of images in which "friends" are pictured in spontaneous moments of fun and adventure. Such scenes are likewise popular motifs among the backpackers interviewed, along with iconic images of Australia. Additionally, Jenkins finds these post-tourists play with notions of reflexivity and tourism marketing by crafting their own "promotional" scenes. These images are then shared with friends and displayed upon returning home. Thus, this research suggests modern attempts at macro-scale social integration continue into the *post*modern era but with a greater emphasis on self-reflexivity. As Haskins (2003) notes, postmodernity does not necessarily preclude the existence

Figure 2.1 Tourist Posing in Front of Machu Picchu, Peru: The Circle of Representation Exemplified (by Nicholas Wise)

of master narratives; the narratives themselves are simply treated differently, with greater skepticism and fragmentation. Indeed, Oakes (2006) argues self-reflexivity is crucial to the transition from modernity to postmodernity. Along with a fragmentation of the grand narratives of modernity, individuals develop rich, layered identities that are built upon through performances of self. Reflexivity—considering the image of one's self among others in society—does not negate the ritual attitude of tourism but does complicate matters. Tourists are prone to reflect upon their own individual experiences and roles, and are concerned not simply with the maintenance or complication of tourism rituals. As such, the ritual attitude of sightseeing continues, but with perhaps slightly different goals and outcomes.

Tourism as Ritual

While both MacCannell and Graburn see tourism as a ritual, their differences in perspective lie in the scales of analysis they employ. As mentioned above, MacCannell's ritual of tourism is limited to the act of sightseeing. It is driven by alienation and a desire for social integration. Graburn, however, considers the broader picture of tourism, situating it in a ritual framework influenced by the work of van Gennep (1909), Durkheim (1912), and Turner (1969) regarding religious rites and rituals, and reports on the experiences of liminality during travel. Liminality in particular is central to theorizing tourism as ritual; as the ritual of the tour progresses, tourists find themselves in varying states of in-betweenness.

Tourism as Liminal

Considering tourism as a ritual, Graburn takes note of the structural elements of tourism, its role in society, and its spheres of experience (profane, liminal, and sacred). He contends, "tourism is one of those necessary structured breaks from ordinary life which characterizes all human societies… [it is] a separation from normal 'instrumental' life and the business of making a living, and offers entry into another kind of moral state in which mental, expressive, and cultural needs come to the fore" (1983, p. 11). He identifies a beginning, middle, and end to tourism as a ritual, with each stage accompanied by specific rites. The tourism ritual begins with the rites of preparation for departure—booking accommodations, scheduling house/pet/child care, packing. The period of time away from home is liminal. In liminality, people, in this case tourists, play with the familiar and defamiliarize it (Turner, 1982). Within the liminal space/time of tourism, one is a tourist, separated from the everyday work/home life—one instead relaxes, plays, adventures, and so on. Rites of touring include exploring new environments, sightseeing, taking photographs, and buying souvenirs. The end of the tourism ritual culminates with the return home, accompanied by rites of re-entry, including unpacking, telling stories of the trip, displaying souvenirs and distributing gifts. Using religious ritual as a framework for tourism, Graburn also ascribes profane space/time to the everyday home life and sacredness to the liminal period of travel. It is important to note, however, that while Graburn, following Turner's early theories of liminality, argued that liminality initiates a reversal of rituals, we do not find that to be the case in tourism. As tourists we may relax some of our mundane rituals, but rarely do we observe a full reversal of habits, behaviors, and ethics. These concepts are put into practice in Chapter 8 as we examine the ritual performances, liminality, and communitas of being a tourist in Florence, Italy.

Tourism serves both a social and personal function. Similar to MacCannell's discussion of sightseeing, tourism generally is a ritual of social solidarity. It is a socially acceptable and expected way to spend leisure time, as well as signifying deeply held values regarding freedom, health, and self-improvement. Leisure time, argues Turner (1982), suggests freedom, that is, freedom *from* institutional obligations and regulated rhythms of work, as well as freedom *to* generate new symbolic worlds and to transcend social structural limitations. Further, leisure time, and tourism more specifically, functions as a way we mark the passage of time. Annual trips and family vacations are repetitive and predictable ways to mark cyclical time (Graburn, 1983). Other, once-in-a-lifetime trips—a graduation excursion, honeymoon, or retirement trip—mark the passage of linear time and thereby function as rites of passage (Graburn, 1983).

However, tourism, in serving a social and personal function, also creates a liminal space/time of action where an individual has an identity outside of that maintained in everyday life. Turner (1982, p. 113) notes, "the liminal period is that time and space betwixt and between… [one] is neither what he [sic] has been nor is what he [sic] will be." In this sense, liminality can also alleviate, if only while on tour, some of our sense of alienation from our ordinary lives. This relief from alienation stems from the liminal nature of tourism and from the way that, once at the destination, we become part of a new community—one potentially quite different from that we left behind. Arrival at the destination inserts the tourist into the community of other tourists. Simply by being at a destination, a bond is formed (for example, all persons visiting Venice) through joint experience of the destination. To the extent that tourists take in the same sites, that bond is furthered. Indeed, in Turner's classic work on ritual (1969), it is liminality that creates community. While theorizing pilgrimage, he observed that the liminality of the collective religious, ritual action produced a space of greater equality in which socioeconomic differences were minimized.

Liminality, thus, has the potential effect of reordering one's position in society and fostering temporary, spontaneous communities, or communitas (Figure 2.2). These processes of liminality and communitas are quite applicable to tourism, an idea not missed by Turner who asserted, "a tourist is half a pilgrim, if a pilgrim is half a tourist" (Turner and Turner, 1978, p. 20). As such, the liminality of tourism offers a space in which the tourist can push back against the rigidness of socioeconomic divisions; after all, the rank-and-file worker and the CEO may sit next to one another at dinner on a cruise. It is the destabilization and reordering liminality provides that allows new communities to spring forth and old orderings to collapse. But as Esposito (2010) has shown, community is a gift and like all gifts, it is something given within the framework of an obligation. It is this obligation that unites us as tourists, that creates communitas, the sense of community. As Turner (1969, p. 97) states, communitas is simply, "[a] matter of giving recognition to an

Figure 2.2 Touristic Communitas in the Dead Sea (by Michelle Israel)

essential and generic human bond, without which there could be *no* society." This sense of the essential and generic human bond is heightened when on tour because we have much in common with our fellow tourist—like her or him we know little about our location, we often must trust the intentions of others, we don't know the best places to eat or drink or how to circumvent the long lines at a popular attraction. In such conditions, that our fellow tourist is from Los Angeles and we are from Indianapolis matters little.

The movement from the profane/everyday to the sacredness/liminality of tourism is accompanied by a series of rituals, collective as well as personal. A common ritual of re-entry to our everyday spaces includes the purchasing and gifting of souvenirs. As such, it is the liminal nature of tourism that suggests the power of souvenirs. As objects that traverse liminal and everyday spaces, they are affective—they are embedded within the tourism experience, conjure sensations, and evoke emotion. Souvenirs, thus, function as objects of memory and metonymy. Morgan and Pritchard's (2005) study of souvenirs offers a rich illustration of how the objects collected during the liminal period of travel function in our everyday spaces. In an autoethnographic examination, the authors take turns narrating the significance and moments of acquisition of some of their most prized souvenirs. From works of art and local crafts to collected mementos and tourism kitsch, the objects they describe are steeped in ritual and memory. The souvenirs are situated in the mundane spaces of home and everyday life, but the memories they evoke are exceptional. The authors tell stories of extraordinary experiences during the liminal period of travel, which are retained in the souvenir's aura. These objects collected during the liminal, sacred space/time of travel infuse their uniqueness into the banality of home and office. Furthermore, souvenirs also act as material means for measuring time, as they reference trips taken and particular periods in one's life. The ritual of tourism can thus be evidenced in the materiality that results from travel experience (Rickly-Boyd, 2012).

Tourism and the Sacred Center

From MacCannell's explication of the ritual attitude of sightseeing to Graburn's investigation of structures of tourism experience, the scope of analysis broadens towards tourism generally. Cohen (1979) further enriches the investigation of tourism and ritual by employing the work of Eliade (1971) and Turner (1973) regarding sacred geographies to theorize the phenomenology of touristic motivation and experience. From Turner (1973), Cohen posits that most individuals reside outside of their sacred, religious centers of meaning, thus necessitating pilgrimage. Because most individuals *do* reside within their social centers (life space), the motivations for tourism—as opposed to pilgrimage—tend to be varying degrees of social alienation. The severity of this social alienation results in touristic pursuits that lie on the following spectrum: recreational → diversionary → experiential → experimental → existential. Thus, like MacCannell, Cohen is relating tourism motivation to alienation; however, unlike MacCannell, he is situating it within a spiritual/existential framework, rather than a Marxist critique.

Cohen's spectrum of varying modes of touristic motivation and experiences corresponds with the tourist on one end (recreational, diversionary) and the pilgrim (existential) on the other. While not using the ritual metaphor explicitly, Cohen does detail the ritualistic functions of each mode of travel for alleviating feelings of alienation. For the recreational tourist, motivation is time away from the stresses of everyday life. Yet, feelings of alienation are minimal and there is a desire to return to the social center. Thus, recreational tourism is more about pleasure; it recreates and rejuvenates. The diversionary tourist, however, experiences a deeper sense of alienation. As such, tourism is a mere escape that makes the alienation of everyday life tolerable. The experiential and experimental touristic modes are driven by a sense of alienation, and tourists therefore seek out authenticity elsewhere. While the experiential tourist merely wants to observe and experience authenticity of the Other momentarily (similar to the function of MacCannell's "staged authenticity") (Figure 2.3), experimental tourists seek to sample various other societies and lifestyles so as to find an authentic way of life for which they might strive upon returning to their social center, or, if that fails, they may take up a life of continual pursuit. At the end of Cohen's spectrum is the existential tourist or postmodern pilgrim whose dire motivations to experience a new way of life leads her or him to commit to a new, elective center, transforming his or her lifestyle or permanently relocating to a new life space. Thus, Cohen provides productive insights into the sociological functions

Figure 2.3 Tourists Watch Hula Performance on Waikiki Beach, Hawaii

of tourism as ritual given varying touristic motivations. What is lacking is attention to grounded touristic performances in tourism places.

For Cohen, tourism becomes a ritual act that is performed as a means of mitigating social alienation in the life space. For recreational and diversionary tourists, the ritual of tourism is regularly performed as a pressure valve to release the stresses of everyday life. The experiential and experimental tourists, however, engage tourism as a ritualistic pursuit of authenticity that necessitates travel outside of their social centers. And, for the existential tourist, tourism becomes the ritual act through which another way of life is entered by thoroughly changing one's lifestyle and maintaining connections through travel to a new spiritual center, or by relocating altogether. While Cohen allows us to extend the role of ritual in tourism to the motivations and destinations of travel, rituals in the perspectives laid out thus far are still removed from the dynamics of the physical spaces of tourism; they do not speak to tourism performances on the ground, in place, from a geographic perspective.

Ritual and the Production of Tourism Place

In taking a ritual perspective on tourism, whether it be the entirety of the act of touring, motivations for travel, or the function of sightseeing, these scholars continually draw from theories in religious studies. The aim of this book, however, is to take a geographic perspective on tourism performances of place. In that spirit, we must begin to transition from theories of social practice to their spatial relations and material manifestations. It is, therefore, worthwhile to (re)turn to religious studies to consider the role of ritual in the production of place, which means, of course, attention to mechanisms that produce *sacred* space. Chidester and Linenthal (1995) note several characteristics of the production of sacred space: sacred space is ritual space; sacred space is significant to the interpretation of what it means to be human, that is, sacred space anchors a worldview in the material world; sacred space is contested. As these aspects of sacred space suggest, sacredness extends well beyond the religious sphere to spaces that are nationally, culturally, and civilly sacred. Additionally, taken alone, each of these characteristics also describes tourism spaces. We begin with the relationship of rituals to the sacred, drawing out metaphorical connections in the (re)production and social significance of tourism places.

In his text, *To Take Place: Toward a Theory of Ritual,* Smith (1987) argues that ritual performance is essential to the consecration of sacred space. Consecration depends not only on ritual but also on the interpretative potential of the site (Chidester and Linenthal, 1995). Sacred space does not erupt out of nowhere as Eliade had theorized, but it is produced through ritual performance and narrative communication and must be continually (re)produced through those practices in order to maintain its significance. More specifically, Chidester and Linenthal (1995) suggest it is through the spatial processes of ritualization, reinterpretation, and contestation that place continues to be reproduced as sacred. In tourism, the processes of site selection and (re)interpretation for tourism audiences is just the beginning of destination marketing (see Chapter 4). A successful tourism destination results in its acceptance by tourists and their ritual performance of the space, the act of traveling to and touring the site (Figure 2.4). Contestation is thus inherent to this process. Tourism boards select and promote "must see" sites from innumerable choices. Sears (1999), for example, traces the development of several early tourism sites in the United States, including Niagara Falls and Mammoth Cave, among others, illustrating how each was reinterpreted as naturally sacred spaces. These early tourist attractions, argues Sears (1999, p.10), "were consumer products, promoted by the railroads which carried tourists to them and organized to satisfy the customers who had the leisure to enjoy them, but their role as products was often marked by religious rhetoric or behavior." Similarly, in a distinctly modern crafting of a national identity narrative, the "See America First" campaigns of the late 1800s to early 1900s were designed to establish "American" tourist attractions, embed these sites in a national tradition, and solidify the American identity through their collective, ritual performance (Shaffer, 2001).

With ritualization and consecration also comes the possibility of desecration. Chidester and Linenthal (1995) argue that it is through ritualized consecration of sacred space that the conditions of desecration are also determined. Defilement, a particular form of desecration, is a violation of ritual order resulting in a loss

Figure 2.4 Tourists at the Martin Luther King, Jr. Memorial, Washington, D.C.

of purity. To counteract this, rites of purification or exclusion can be utilized to remove the "polluting" source. In the case of most tourism sites, it is the processes of reinterpretation, inversion, and hybridization that are put to use. These processes are prolific at tourism sites as they allow for multiple narratives in a single space. According to Chidester and Linenthal (1995, p. 19), inversion and hybridization, particularly, lend themselves to "the kinds of 'desecration' that symbolize alternative relationships to sacred space." Foote's (1997) research into the processes of commemoration following acts of tragedy and violence illuminates these dynamics. In theorizing how events are marked, rewritten, ignored or even obliterated from the landscape following traumatic acts, he draws from a diverse set of cases, including sites of Texas state history (the Alamo, San Jacinto, and Goliad), the Mormon exodus, and the Salem witch trials. However, by also interrogating sites of mass murder and accidental tragedy, Foote reveals that commemoration processes can invert and hybridize tragedy as a source of community cohesion. Thus, many such sites can be included among "dark" tourism destinations (for a critique of the theorization and language surrounding dark tourism, see Bowman and Pezzullo, 2009). But such processes are not limited to historical sites; rites of purification and exclusion are central to all-inclusive tourism resorts (Figures 2.5 and 2.6) and theme parks as well. Indeed, engaging theories of the mechanisms of the (re)production of sacred spaces offers much for future tourism research.

Figure 2.5 Private Property Sign: Couples Swept Away Resort, Negril, Jamaica (by Paul Kingsbury)

Figure 2.6 Sandals Resort Security Guard, Negril, Jamaica (by Paul Kingsbury)

Weaving through these discussions of sacred space are both the politics and poetics of space, that is, the production as well as experience of space. Considering theories of sacred space both in and outside religious studies, Chidester and

Linenthal distinguish two lines of definition regarding the sacred—substantial and situational. Substantial definitions (Otto's "holy," Leeuw's "power," Eliade's "real") highlight the manifestation of sacred spaces in the ordinary/profane, and as such are primarily concerned with the poetics of sacred space. Situational definitions (Durkheim, van Gennep, Levi-Strauss) emphasize the practices and politics of sacred space, and are, therefore, most useful for tourism studies. While substantial definitions of sacred space may be worthwhile for understanding the tourist experience, situational definitions offer a relational approach for integrating ideology, semiotics, and tourist performances, as well as for the simultaneous performances of the sacred and the profane in tourism places. Rather than separating definitions of sacred space into respective spheres, attention to the performance of tourism place engages both perspectives. In fact, tourism performances necessitate both the politics and poetics of space, as much because most tourism sites are not singular in purpose or interpretation. Indeed, tourism sites are the everyday spaces for those who live and work in and near them. Yet, they function as sacred, liminal spaces for tourists. Such dissonance in tourism experience is evidenced when one bumps up against poverty in paradise, for example.

Myriad rituals exist to convince economically privileged tourists that they are not, in fact, confronting poverty while on vacation. As Kingsbury (2011) notes, however, scholars of critical tourism studies could be more nuanced in their analyses of disparities between wealthy tourists who seek paradise and the often marginalized, commodified local laborers who serve them. Advancing a psychoanalytic argument in an organizational ethnography of Sandals resorts in Jamaica, Kingsbury notes a complex web of sublimation and enjoyment fuels many interactions between tourists and workers. In company literature, marketing materials, and everyday interactions, the enjoyment of resort workers is offered as both a substitute and addition to the joy of patrons. "Smile awards" are given to employees who visibly cause tourists to enjoy their stays the most (e.g. prompting laughter), and in Sandals' guidebooks, workers are always portrayed as exuberant. A quietly ominous sign greets workers at Sandals Negril—it reads: "Remember the steam kettle; though up to its neck in hot waters, it continues to sing" (Kingsbury, 2011, p. 651). As Kingsbury's example demonstrates, problematic rituals of tourism that reify poverty and privilege manifest in a diversity of situations. Nevertheless, in this case it is the performance of such problematic rituals that continually constitute place.

Conclusion

As we have noted in this chapter, considering the ritual aspects of tourism was a precursor to the contemporary studies of tourism performances. MacCannell's assessment of ritual in tourism was limited to the act of sightseeing. Tourism, however, does not end at the destination but in the return home, with memories and materialities of travel that have potential for continuing these experiences.

It is with this in mind that Graburn developed his theory of tourism as a secular ritual, and in so doing we can see MacCannell's theories of sightseeing nested within a broader ritual act. Cohen further considers the recreational to existential motivations of participating in tourism as a ritual. Yet, there remains additional room in tourism studies for the integration of theories of ritual, particularly as they relate to the social construction of place. Taking a geographic perspective on the relationship of the tourist to the site calls forth attention to social practices that produce space. Tourism rituals do much more than offer the potential for social integration; their continual performance reifies tourism places.

Rituals of tourism are paradoxically upheld and broken at every turn. While tourists across many places perform very similar rituals, as we have argued above, ritual acts also (re)present an opportunity to break tradition. Moving from notions of a coherent modernity to late- or post-modernity, ritual experiences have the potential to become more fragmented and contested. Digital and social media are transforming the ways in which touring is documented; tourists increasingly remain more connected to "home" when they would have previously inhabited more distinctly liminal spaces. Despite these shifts, however, the ritualistic structure of tourism, while malleable, remains recognizable.

In the following chapter we move to consider the communicative devices of tourism rituals, signs, and the processes by which tourists interpret them. By focusing on a Peircean semiotic, in particular, we stay attuned to the materiality of place in tourism performances and tourist meaning-making. Ritual, however, is also a foundational tenet of Chapter 4 on ideology and tourism sites—marketing, promotion, and interpretative materials. Ideology does not simply exist; it is produced through social practice. It must be recognized and reified in tourism spaces. In Chapter 5 we turn back to semiotics and ideology, as well as build from the earlier theories of ritual presented in this chapter, to argue for a performative approach to tourism as a means to cohesively consider the tourist/site relationship.

Chapter 3
Semiotics and Tourist Meaning-Making

Introduction

Within the geography literature, place is understood as space made meaningful. Through interaction and experience with and in place, space is infused with meaning, emotion, and significance (see Chapter 1). Tourism is fundamentally a place-based endeavor in which the term "destination" underscores the importance that locale has for the very act of touring. Being in place, tourists must make sense of the landscapes in which they find themselves, and in so doing, their experiences of place support or alter their preconceptions. One of the critical questions regarding this sense-making process is the role of the tourist vis-à-vis representation and materiality, or in other words, how tourists create meaning from images and experiences of place. While the previous chapter concerned the ritual aspects of tourism as a social practice that both motivates tourists and sacralizes sites as tourism places, this chapter explores the intersection of the tourist and the site as an experience of place.

Places are made meaningful by interpretation, an active process that involves objects and interpreters (Houser, 1992). In his work on semiotics, American Pragmatist Charles Sanders Peirce wrote a great deal about the process of how objects and interpreters co-construct meaning within specific contexts. The object provides constraints to interpretation as does what Peirce referred to as collateral information—the accumulated notions, concepts and information we have previously gained about the object and objects like it. Both, in turn, shape the interpretation of the object. With regard to tourism, the crafting of place occurs at multiple moments both before and after arriving on site, from guidebooks and tourism brochures to the physical encounters with tourist landscapes. This process continues to evolve upon return with the retelling of tales and the material objects and images (photographs, both printed and digital) that accompany the tourist back home.

In tourism, place-making hinges on tourists' abilities to create meaning before, during, and after their travels. In order to highlight this process, this chapter will unpack the insights that Peircean semiotic theory offers to understanding more fully the processes of interpretation *in situ*. While semiotic studies tend to be based in the French school of Saussurian semiotics, the coeval American version of semiotics based on the Pragmatist tradition offers a more robust way of dealing with the materiality of objects by moving us beyond linguistic associations. We begin with Peirce's notions of habit, which are relevant to shifting our focus towards the tourist in place. The interpretive process, while brought more explicitly to the fore when we are on tour in a strange locale, is nevertheless the default mode for engaging with the world around us, even when we are in familiar environs and involved in our everyday lives. Indeed, reading landscape as a text is something we necessarily do on a daily basis; thus, when on tour we apply this practice to tourism sites. Taking this a step further, we then consider tourism as a case of embodied semiotics in which tourists do not merely read their surroundings but perform tourism. Finally, we attend to intersubjective and co-constructive processes of meaning making. As tourists we do not decipher the meaning of places alone, in our own minds, but through interaction with other tourists, locals, guides, and guidebooks. As semioticians our work is never complete; it is an ongoing, dynamical process.

Habits of Interpretation—A Brief Overview of Semiotic Theory

This section provides a short introduction to semiotics and to the ways in which representation and interpretive habits play out in tourism. It will offer a brief version of Peirce's three-termed process in order to highlight the role that objects play in guiding our interpretations and the role that collateral information plays in shaping our meaning-making. This section provides the theoretical foundation for exploration in the later parts of the chapter of the ways in which embodied experiences occur (and are (re)interpreted). To that end, the idea of habit will also be addressed with particular attention to our banal activities and daily rituals as actions that do not stop when we are on tour but become sharpened and more explicitly grounded in deciphering/interpreting our surroundings.

Signs

Semiotics is most easily defined as the study of signs and the interpretation of their meaning (Nöth, 1990). While we may think literally about actual signs (such as traffic signals, signage, and placards), language systems function as a kind of sign, as do gestures, clothing, sounds, physical conditions such as fever, and natural phenomena such as clouds. As such, a sign is anything that conveys meaning. The world is perfused with signs and so semiotics can also be seen as a kind of epistemological process whereby we make sense of the world around us (EP 1.30[1]).

While semiotics can be traced back to the Classical World, modern semiotics came to the fore in the late nineteenth and early twentieth centuries on two opposite sides of the Atlantic Ocean. Working in France, the Swiss linguist Ferdinand

1 In following the standard Peirce citation method, we reference his scholarship by the work (each are noted with two capital letters CP (volumes 1–8), EP (vols. 1–2), and SS), the volume followed by colon, followed by page numbers (e.g. EP 2.478) unless it is SS, which just includes the work and the page number.

de Saussure (1983 [1959]) was seeking to understand the underlying structural relationship between language and meaning. At the same time, in the United States, scientist and philosopher Charles Sanders Peirce (1992–1998, Hookaway, 1992; and Friedman, 2004) was developing theories about signs and the process of inquiry, and working for the U.S. Coastal Survey.

Saussure (1983) proposed a two-part sign composed of the signifier, or the *image acoustique*, and the signified, or the concept to which the sound referred. Essentially, Saussure was interested in the relationship between sounds and the ideas they generate. He observed that a fundamental aspect of this relationship is the arbitrary nature of associations. The idea of a furry, four-legged creature that barks (the signified) is only linked to the spoken word "dog," *hund*, *perro*, *kutya*, or *inu* (the signifier) by agreed upon habit or custom. Once a signifier is set, however, the term can accrue both denotative and connotative meaning, but in both cases the meanings are dependent upon common and collective usage, not on an intrinsic linkage between the idea and the sound. Equally important with respect to Saussure's work is the structuralist nature of his investigations into grammar and the way in which meaning emerges from the ordering of language. Each language is ordered and referential in regards to the place of the sign. "Language," he wrote, "is a system of interdependent terms in which the value of each term results solely from the simultaneous presence of the others" (Saussure, 1983 [1959], p. 14).

Saussure's structuralist model of semiotics was influential in the so-called linguistic turn (Giddens and Turner, 1988) and has become the more commonly known theoretical perspective of semiotics. This linguistic basis for Saussurian semiotics has, however, been ill equipped for applications of sign interpretation outside of the realm of language to areas of materiality or to habits or actions, and thereby tourism. As Nöth (1990) writes in his survey of semiotics, "Saussure's contribution to a general theory of signs has been only minor. He had little to say about nonlinguistic signs and was not concerned with questions such as the general typology of signs" (63). Nevertheless, semiotics as taken up in many other disciplines, tourism studies included, has often taken for granted that "semiotic" is a synonym for "symbolic." While producing important discussions of the role of the symbolic in tourism, these studies do not engage with the ways in which meaning is educed from objects such as architecture, lampposts, and street signs—or the ways in which the physical world and our habits and actions are examples of semiotics in action (Culler, 1981; Davis, 2005; Frow, 1991; Gaffey, 2004; Jaworski and Pritchard, 2005; Nelson, 2005; and Smith, 2005). Peirce's theory moves us beyond simply talking about symbolic meaning to explore much more explicitly epistemological and ontological questions of the processes through which we understand and are in the world around us (Freidman, 2004; and Liszka, 1996).

Additionally, Peirce recognized the arbitrary and rather symbolic relation between object and interpretant (CP 2.92 4.447; EP 2.5–9), but unlike Saussure, he also recognized that signs might relate to their object in other, more determined, less contingent ways. For example, signs might be indices and be in actual relation—a fever as an index of infection or a paw print as an index of an animal in the woods

(CP 2.304; CP 4.447; EP 2.163; EP 2.274). Signs might also be iconic, that is, they might be likenesses of the object or they might share in its essence, such as a map (CP 8.335; EP 1.226; EP 2.5). While Peirce spent a great deal of time on typologies (EP 2.296; 2.481), he conceded that rather than stark divides between the ways in which signs relate to their objects, they more likely share in the indexical, iconic, and symbolic relationship with some aspects being more prominent than others (EP 2.10). Within the context of tourism we can identify examples of each. Monuments, Peirce contended in discussing the one at Bunker Hill (EP 2.163), are indexical; they draw attention and like a pointing finger, they say "here." Food can function iconically in that the foodways of a place have developed not without reason but because of the agricultural and socio/cultural practices tied to the place and community. Lastly, signs can be symbolic, for example the throwing of coins into fountains for luck.

Peirce's Theory of Signs

Peirce's writing on semiotics was both at the center of his philosophical thought and developed in tandem with other theoretical interests. In particular, and as discussed below, perhaps the most telling difference between the European linguistic version of Saussure's semiotics and the American version, is that Peirce's theory was intricately linked to his pragmatism (Short, 2004). At present his collected works span eight volumes and while there are specific essays and lectures that explicitly treat his semiotics, insights are also gleaned from across the collection of his works (Peirce, CP 1–8). Peirce's ideas morphed and evolved over his lifetime, sometimes resulting in major changes and other times in new terminology (Liska, 1996; and Friedman, 2004). Generally, Peirce scholars can identify the outlines of his theory and the key areas that remained constant. For the sake of simplicity, what follows is a basic outline of Peirce's theory (for a much more nuanced look, see Metro-Roland, 2011; also Friedman, 2004; Hookway, 1992; and Liszka, 1996).

Peirce identified three aspects of a sign: the *object*, the *representamen*, and the *interpretant* (EP 2.272–73). Confusingly, Peirce uses the term *sign* to refer to both the whole triadic unit as well as the middle term, the representamen, but this interchangeability also allows us to see more clearly the way this entity functions cohesively. As a three-part sign—if we pull apart the various terms that Peirce uses—it consists of an object, something that is tangible (for example, a bicycle or a shout), a thought or image of it, the representamen, and an interpretation (which can be an understanding or an action) which Peirce referred to as the interpretant (EP 2.272–73; EP 2.478). One way to explicate this relationship is to turn to the idea of pragmatism. As mentioned above, Peirce's semiotic was intricately linked to his development of this quintessential American philosophy. The Pragmatic Maxim as stated by Peirce is: "consider what effects, which might conceivably have practical bearings, we conceive the object of our conception to have. Then our conception of these effects is the whole of our conception of the Object" (EP 1.132). It may seem that interpretation lies solely with the interpreter and that this allows for any interpretation, even wildly inaccurate ones. Peirce, as a working scientist, was invested in the notion of fallibilism—any interpretation is contingent, always open to further tweaking (EP 2.499). As Nathan Houser (1992), the contemporary editor of Peirce's work, writes

> The interpretant is, or helps make up, a habit that 'guides' our future (and present) actions or thought with respect to the object in question, or objects *like* the one in question. If the interpretant is untrue to the object, our behavior will not be (or *may* not be) successful—reality will have its way with us. Not until our interpretations (our ideas or intellectual habits) are fully attuned to their objects will we avoid unexpected confrontation with a resistant reality. In this way, the real object determines or shapes our mind, our reservoir of intellectual habits. (EP xxxix-xl)

In other words, we as interpreters do not have free reign since the material world will get in the way. As a result, our interpretations are never complete, but continually open to reassessment, conceptually as well as in the actions that result from them.

Habit

Whereas Saussure focused on the relationship between words and ideas, Peirce was very interested in the ways in which interpretations manifest themselves in concrete action, or habit (EP 1.129). To illustrate, let us imagine a boy and a girl in high school dating, and the girl has an uncommon diminutive with which she addresses her love. And let us say the couple grow up, go away to college, and part ways, and he finds a whole new set of friends who know nothing of his pet name. Yet one day as he is walking down the street, he hears the name called out and instinctively reacts to the sound, looking around to see who said it, perhaps feeling hopeful, perhaps feeling anxious. It is a word, a signifier, but it elicits a reaction. Not all objects have a singular interpretation or reaction, but this example does highlight the ways in which signification goes beyond the Saussurian signified to consider how certain objects and tangible things evoke embodied response or what Peirce termed "habit."

Indeed, many interpretations lead us to perform in a certain way (Edensor, 2000; Bærenholdt et al., 2004; and Rickly-Boyd and Metro-Roland, 2010). Crucially, this performance is not merely cognitive, but rather a reaction, a performance of habit in the Peircean sense. That is to say, a habit engrained in the body that is subconscious still puts bodies in motion. Furthermore, a habit is not of the object, but of experience where that experience is both learned and practiced, where it is socially produced by the interplay of agency and social structures in a process of identity construction. It is in this way that notions of practice and embodiment are embedded in Peirce's semiotic. But such habits do not necessarily always serve us well, as Houser (1992) points out. They may place us in danger; they may lead to inappropriate actions; they may not resolve situations. Confronted with a moment of danger, we must pause and adjust our interpretation. Such moments foreground the relationship between interpretation and collateral information.

Collateral Information and the Dynamic Interpretant

Interpretation, Peirce notes, does not happen in a vacuum. We are not *tabulae rasae* on which we must start from scratch in each interpretive exercise. Rather, we bring to bear what Peirce at some times terms "collateral observation" and at others "collateral information" (EP 2.493–96). This is the collection of experiences and information we employ, sometimes explicitly but often implicitly to make sense of objects we confront. But this collateral information is fallible and interpretation is always subject to negotiation and revision (CP 1.171). That interpretation is always open to revision can be seen from the fact that Peirce accounted for the *longue durée* of interpretation, of changing cultural norms, of mistaken notions along the way in dissecting the three elements that constitute his sign into ever finer categories: of immediate and dynamic object and immediate, dynamic, and final interpretant (EP 2.478; SS 111). For our purposes, one may think of the immediate interpretant as a "first impression," the dynamic interpretant as a series of interpretations that evolve over time and the final interpretant as a final resolution of meaning. Thus the Peircean notion of the interpretant is, as Eco (1976) has noted, ever receding. It is never totally settled or final, but rather continually dynamic. This must be so because any discussion of a final interpretant implies that collateral information has reached some point with respect to an object wherein it cannot become greater. Short of death, collateral information always has the potential (often unmet) to increase.

If, for example, we come from a culture in which white is associated with death, we may be perplexed to find it employed so overtly in celebration of a joyous occasion, but our recognition of other clues would help recalibrate our interpretations. Color is an example of a sign that is culturally contingent. There is no a priori link between white and joy or sorrow, but once the cultural parameters are established it is left to the iconoclast to break them. Yet anyone who has spent time in both the U.S. and Japan, and thereby relies on one's collateral information of these contexts, is aware of the opposite associations of white in the two countries and would act accordingly.

This last example is also instructive because it illustrates that collateral information also is at least partly held in common, particularly as it is used to inform collective identities—identities of family, religion, region, and nation. It is this commonly held information that is in many ways the most complex because it is a function of the play between individual agency and societal structure (see Althusser, 1971; Giddens, 1984; and Žižek, 1989). This last point is crucial because it points to the many ways in which collateral information contains within it various scales of ideology. As is taken up in the following chapter, post-

Lacanian ideology theory, be it Althusser or more recently Žižek, suggests what hails us is some sort of "sublime object" (Žižek, 1989), or what we might label an aestheticized Peircean object, while our performance as ideological subjects arises out of exactly the sort of habit envisioned by Peirce. Put another way, within the ideological framework of our society, there are a large number of things that have more or less agreed upon, and therefore taken-for-granted, meanings. The processes by which we learn these collective ideologies are termed interpellation, which functions socially to continually reinforce shared meanings but also both individual and collective embodied reactions, or habits (Althusser, 1971; Žižek, 1989; and Eagleton, 1991). By way of examples, consider traffic flows (driving on the left or right hand side of the road), holiday observances, and reactions to national iconography.

Semiotics and Everyday Life

Peirce's semiotic theory would argue reading the landscape is something we all do; otherwise we would not be able to function in the world (EP 1.30). In moving through the world, we must make sense of what is around us. To offer another analogy, even if we do not have the horticultural training to identify the various plants and flowers in a garden and to see that the gardener has planted native species that attract monarchs, we can still enjoy the butterflies and the flowers.

While the interpretive process may begin with the object, it does not rely solely on the object itself but draws upon a whole scaffolding of knowledge and experience to move interpretation forward. Just as novels, poems, and essays are not single entities, unique and independent unto themselves, but are embedded in larger webs of critique, historical context, authorial biography, similar and dissimilar works, and genres (Eagleton, 1996), the landscapes produced by and for tourism are similar to texts by virtue of the fact that they can be created, cultivated, shaped, nurtured, and manipulated to support particular interpretations.

The interpretive theory posed by Peirce, which did not treat literature *per se* but epistemology more generally, offers a helpful guide in our understanding of landscape interpretation as it makes clear that "reading" a landscape is an intersubjective process between the object and the interpreter. But more importantly and counter to Lewis (1979), reading the landscape is something that people do whether they are necessarily conscious of it or not.

As the humanistic geographer Yi-Fu Tuan (1974) has argued, while there are underlying cultural norms we bring to our understanding of a landscape, these interpretations can also vary according to context and personal histories. As such, Tuan juxtaposed "public symbols" against "fields of care." The former are intentionally created or delineated to be understood from the outside, whereas the latter category refers to spaces that are understood from the inside (and they might in fact be inaccessible or undifferentiated to those on the outside). Because interpretation does not take place in a vacuum and we cannot easily ignore our experiences, the collateral information we bring to a place is not easily overridden. Even in the case of public symbols, there may be deeper relations to the place, turning the site into a field of care. Any visit to the Vietnam Veterans Memorial in Washington, D.C., a public symbol, demonstrates the ways in which this public monument resonates with personal feeling for those who have loved ones or friends listed on the wall.

Thus, one is not just a semiotician when on tour. We all get a great deal of practice as we go about our everyday lives. Within the realm of the natural and built environments, place-making and place interpretation are almost universal occurrences. Landscape geographers, while not necessarily approaching their studies from the point of view of semiotics, highlight the ways in which the worlds we inhabit are not just empty spaces or meaningless expanses but that the forms, features, colors, and contexts of the environment around us have value, import, and significance (Meinig, 1979; Rowntree, 1996; Groth and Bressi, 1997; and Wilson and Groth, 2003; and Cosgrove, 1984). Similarly, in the context of semiotics, there are no empty signifiers in the built or natural environments.

Tourism as an Embodied, Semiotic Process

As noted in the previous chapter, Dean MacCannell was among the first tourism scholars to take notice of the semiotics of tourism, and sightseeing in particular. MacCannell (1999, p. 41) argues that a tourist attraction is defined as an "empirical relationship between a tourist, a sight, and a marker (a piece of information about a sight)." The marker is an essential first step in the construction of an attraction. In providing information, it sets the object apart from similar objects as worthy of attention. A Peircean framework was thus influential as MacCannell developed his theory of tourism attractions as signs. While a strictly Saussurian semiotic would limit his elucidation of the tourist/attraction relationship to a conceptual one, a Peircean semiotic, he argues, allows for the consideration of tourist behavior. This is particularly important given MacCannell's observation of the "ritual attitude" of sightseeing (see Chapter 2). Sightseeing, he contends, "is a ritual performed to the differentiation of society" (1999, p. 13). While differentiations are the attractions of tourism, he suggests tourists perform the modern ritual of sightseeing as a means towards social integration, a way to overcome the discontinuity of modernity—as he notes, "the act of sightseeing is a kind of involvement with social appearances that helps the person to construct totalities from his [sic] disparate experiences" (1999, p. 15).

One can see how being on tour can stretch the semiotic process. One's interpretation of the site and hence behavior within it falls under Peirce's notion of the interpretant and the habits of behavior that such interpretation yields. Much of the time while on tour, our existing collateral information may serve us well or good enough and the resulting interpretation and habit (thought/action) suffices. It is only when an encountered object "pushes up against the interpretant" (Metro-Roland, 2009, p. 277) that a refiguring of meaning occurs. The subtlety of Peirce's idea of the dynamical interpretant is that after an initial interpretation is made, it is then progressively refined. Consider the case of a misinterpretation that puts the body in motion—in this instance in the wrong direction. Two Americans walk into a British pub and take a seat at a table, imagining that a server will approach to take their order. After several minutes of watching other patrons go to the bar to order drinks, one of the Americans does the same. The next day, when the same two Americans enter another pub, they walk immediately to the bar, place an order, and then find seats with drinks in hand. Such a first attempt at interpretation may solicit a confirmation or rebuttal or rebuke; in this instance, the first attempt at interpretation and action were met with no response, suggesting an inaccurate interpretation had been made. Witnessing others offered an alternative interpretation and appropriate action. This process of "truth marking" (MacCannell, 1976), as an affirmation or refutation of one's actions, allows us to add to our accumulated knowledge of the world. In these instances, faced with new cultural milieu, much that once seemed solid begins to shift and morph.

Tour guides and guidebooks have worked on the premise that they can bring interpretive order to what appears to be chaos by providing the needed collateral information (Franklin, 2004; Koshar, 1998; and Travlou, 2002) (Figures 3.1 and 3.2). But even a well-equipped tourist has moments of uncontrolled and unplanned encounters, and at these times, the interpretive processes that underlie our daily lives but which at home may seem to run on autopilot come to the fore. It is also important to consider that the performative aspects of being a tourist add additional layers of complexity to the ways in which interpretation is undertaken. This leads to thinking about the double act of semiosis, of the ways in which the places of tourism are meaningful and the ways in which the embodied experience of being in place entails a semiotic exercise of interpretation.

Tourists arrive with sets of preconceived expectations from the guide materials and/or other cultural documents which have served to prepare the individual, but they must constantly renegotiate these expectations about a site once there. Whatever

Figure 3.1 **"Danger" at Yellowstone National Park (U.S.) (by Isaac Rooks)**

Figure 3.2 Tourists Using Walkways in Yellowstone National Park (by Isaac Rooks)

they might have read in the guidebook or online, the interpretive process cannot be put aside once tourists have actually gotten into the field (Dann, 2002; Selwyn, 1996; Young, 1999; and McGregor, 2000). Graburn's theorization of the ritual of tourism is a valuable way of thinking about the process, generally; however, we do not find a complete ritual reversal once in the liminality of the tour (see Chapters 2 and 8). Within the context of being on tour there may be a relaxation of norms. For example, while it is perfectly acceptable to wear flip flops and a floral print wrap in Waikiki, the same outfit would likely not be considered appropriate office attire. People sleep in, they may forget their exercise routines, and often times eat and drink to excess. Tourism allows the tourist to enter into a realm outside of that of the quotidian. But being a tourist is not risk free, and the unfamiliar poses a particular challenge: the need to actively interpret unfamiliarity, even for the mundane rituals we bring along with us.

In many ways, it is in the context of tourism that we are most aware of the ways in which we undertake semiosis precisely because even that which is the same is at least vaguely unfamiliar. One may eat a deli sandwich regularly for lunch in the office, but doing so on the beach makes this quotidian ritual rather unique. Thus, one of the ironies of tourism is that although we may escape the work-a-day world, we continue to have to undertake the daily care of ourselves: sleeping, eating, hydrating, defecating, and navigating from point A to point B. While these seemingly mundane tasks should be easy to undertake compared to grasping the nuanced distinctions between Moorish and Gothic arches, for example, often these rather banal tasks prove far more challenging and worrisome.

Unless we take all our meals in the hotel, feeding ourselves—one of the most basic of everyday tasks we carry out—must be undertaken more conscientiously than normal. First there is the need to find a place to procure food. There are the guidebook recommendations, which we must physically locate in the landscape, and there are also the cafés one encounters while exploring. What might the food on offer be? Is this a place that, as a tourist, I might feel comfortable? Can I figure out the menu if it is not in my native tongue? Is this something I want to eat? What is the probability that I am going to get sick if I eat here? As such, the tourist is faced not only with interpreting the signs encountered in the landscape and tourist materials, but also with the task of pairing culinary desires with what might be available. The humorist David Sedaris talks about living in Paris and frequenting the neighborhood hardware store, in spite of the fact that he did not really need any tools and such, simply because the store employees were nice to him and put up with his poor French. The fantasy of going to the truly local café where tourists do not venture is often shattered by the reality of finding such a place—feeling like an intruder because of annoyed/hostile staff and patrons, and eating something awful because the menu was unintelligible. While most gourmands would agree that establishments that cater to tourists are rarely exemplary representations of local cuisine, nevertheless a perusal of any tourist district will find these places well attended because they make things easy for the outsider—they have menus in view, often multilingual ones, they may advertise special tourist menus or regional/national specialties, the kinds of things that locals might be bored with but which are attractions for the tourist. All of these things are signs, in the semiotic sense, which hail the tourist. For the tourist in an unfamiliar place this is a very rational interpretive calculus.

The semiosis of tourism is embodied in the attempts to fulfill bodily needs, but also in the significance of our bodies' presence in tourism spaces. Tourists during their sightseeing are often moving through rather circumscribed spaces, a kind of "tourist bubble" (see Chapter 8), which might be more or less segregated from the actual daily spaces of the place (Hayllar, Griffin, and Edwards, 2008). In the ideal tourist space there are enough locals to give a sense of authenticity (MacCannell, 1976; 1999). So while as a tourist, one sees crowds as an endorsement of a site, and especially if the crowds happen to appear to be other travelers, identifiable by their guidebooks, cameras, backpacks, and casual clothing, the very presence of tourists taints the space for locals or those who prefer not to identify as tourists. Arthur Philip's (2002) novel *Prague* perhaps best underscores the dilemma:

> As early as July 1990 A Házam danced on the precipice of overpopularity; everyone felt that their secret had slipped out of their control. The very hippest Hungarians felt there were too many foreigners. The very hippest foreigners had the impression there were too many uncool foreigners. The rest of the

> foreigners, unaware they were uncool were noticing too many obvious tourists. By September it would become a favorite bar from the past that you couldn't really go to anymore without aching for the good old days when it was yours alone. But for a few weeks in July of that year, before it won praise in a college–published budget travel guide for its authenticity as a locals' hangout, A Házam was everyone's first choice. (p. 79)

The embodied semiotics of the interpretive process of touring plays out in the process of actual sightseeing as well, particularly in regards to the ways in which tourists' behaviors are informed (or not) by tourism spaces. Mosques and churches offer interesting cases of the vagaries of interpretation. While the expected behaviors, or as Peirce would say, the habits that one obtains, for behaving in a church are similar to those for behaving in a museum, there are nevertheless crucial differences. Since religious buildings can today function as either museums, sites of worship, or both, one's interpretation of the site (and hence behavior) undertaken within the site hinges upon how one identifies the site: as museum/cultural site or church/mosque. Interpretation, as we have seen above, depends upon collateral information that we bring to the process and the ways in which the object provides guidelines for our efforts. To reiterate the potential of a Peircean semiotic, whereas Saussure's semiotics would conclude with the concept induced upon encountering the object, Peirce was pragmatically concerned with the resulting behavior that comes from the process of interpretation.

Taking, for example, the case of the Blue Mosque and *Hagia Sophia* in Istanbul, the observant tourist would not enter either and willy-nilly determine that either place is a *hamam* (public bath) or a *bazaar*. But the historical trajectory of religion and culture within Turkey do allow museum/cultural site or church/mosque as interpretative possibilities. One's interpretation of the site and hence behavior within it falls under what Peirce termed the "dynamical interpretant" and the habits of behavior that such interpretation yields. Determining that the Blue Mosque (Figure 3.3) is a functioning site of worship that requires the respectful removal of shoes and the covering of women's heads, while *Hagia Sophia*, as a former church/mosque-turned-museum does not, is a semiotic exercise. The interpretive process is aided by collateral information from guidebooks, tour guides, other experiences with museums and/or mosques, and the immediate clues from those working at both sites—selling entry tickets at one and offering headscarves and bags for shoes at the other. Collateral information notwithstanding, the possibilities of misinterpretation result in female tourists covering their heads as they enter *Hagia Sophia* and some visitors praying at the green marble circle which marks the spot reserved for the Byzantine Empresses in the gallery.

In some cases, the fact that a place has become a tourist site becomes the overriding and dominant interpretive fact, and shapes behavior. The Sistine Chapel has become for most intents and purposes a tourist magnet. It is rare that images from the chapel focus on the entire space, but rather the individual scenes have been served up on postcards, advertisements, umbrellas, posters, and all

Figure 3.3 Blue Mosque, Istanbul, Turkey

manner of quotidian objects. The vast majority of visitors have been seduced by the interpretation of the site as an object of tourist interest, so that even for the faithful, the reverence and quiet one might assume would accompany religious places or even museums, is given over so that the site becomes like the Spanish Steps or the Trevi Fountain: a "must see" tourism site, with a cacophony of visitor noise that irks the guards who still understand the site as a spiritual place, and the art critics who understand the site as a fragile cultural artifact (Knight, 2012; see also Bremer, 2004, for example).

Semiosis as Co-Construction

Semiosis is a co-constructive process, always dynamical, between the object/site and the interpreter/tourist. This is because a tourism site is rarely simply *sui generis*. It is of a type, even when some of the constituent elements are unique. The meaning of a site does not emerge fully formed from the site alone but webs of information and images give shape just as much as the actual physical elements of the site. Paris is, for example, the Louvre, Eiffel Tower, Notre Dame, and the Seine, but it is also Robert Doisneau's photos, the paintings and posters of Toulouse-Lautrec, the writing of Balzac, Zola, Flaubert, Hemingway, Miller, de Beauvoir, and Sartre, as well as Jean-Luc Godard's films. It is the City of Love, the City of Light, but more generically it is a "romantic European city," filled with sidewalk cafés, wide boulevards, wine, coffee, museums, and bookstores. Paris does not have to sell itself extensively because it is over-determined by the representations of it. Vilnius, on the other hand, has more work to do in filling in gaps for the average tourist imagining the place (for more on tourism marketing, see Chapters 4 and 7).

Shaping Attractions

Within artistic production, the concept of genres is the employment of expected devices for shaping the form of a creative work and these in turn shape the expectations of the audience. Genres provide constraints, establishing parameters for production and for reception or understanding. We would not judge a sonnet successful as a sonnet by the same parameters as a haiku, nor an orchestral symphony by those of bluegrass. This does not mean that theories of larger overarching aesthetic judgments cannot be applied, it simply means that within genres there are rules, criteria for assessing.

Because there is so much competition for tourist attention, there is the need for deliberate tourist sites to differentiate themselves and make themselves attractive to visitors who have many options for spending their time. There are two diametrically opposed ways of approaching this marketing: highlighting what is unique about the site, and fitting oneself into a trope, thus helping make the site legible. Rural tourism in the United States, for example, draws on a host of myths (Hopkins, 1998) so that while any particular instantiation of a pumpkin patch or apple orchard might have unique features, images of the rural and the American fascination with farming as a simple way of life, filled with tasty baked goods and jams and uncorrupted by complicated urban politics, also helps shape the expectations of the visitors and the crafting of the site resulting in similarity across places. Best practices in the trade of creating various types of tourism, from agricultural to heritage sites, also work to create a common set of elements which contribute to satisfactory interpretation (Ashworth and Tunbridge, 2000; and Agricultural Marketing Resource Center, n.d.). The historic home converted into a small period museum and local "historical" festivals are found in rural areas around the world (Fehér and Kóródi, 2008). Such small museums function synecdochically as parts of the entirety of (rural and/or culturally specific) history. Is your small town known for the production of blueberries or garlic? Does your community have a famous native son or daughter? From an economic perspective, it makes sense for tourism sites to draw upon already established practices and well-trodden successful formulae. On Oahu, the tourist is offered a luau along with hula dancers, while in Seattle, the tourist is offered a Native American Salmon bake, Coast Salish storytelling, and Native American dances.

As Hoelscher (1998) noted in his study of New Glarus "Little Switzerland," Wisconsin, efforts aimed at conforming to tourists' expectations often end up with other-directed landscapes rather than being reflective of what was there. In this case, the local architecture built by actual Swiss immigrants was further "Swissified"; the same evolution is seen in the history of Solvang in California (Phillips, 2002). The landscape is shaped to fit tourists' perceptions of a general idea of Swiss-ness or Danish-ness rather than any actual exemplar. Indeed, theming, as associated with tourism, has become pervasive (Gottdiener, 2001), resulting in the "touristization" of the locale at multiple scales.

For sites that have a variety of attractions to offer visitors, tourism districts can emerge. Yet, these districts offer many of the same sites and services one might find across the country if not across the world: a market hall, mass produced souvenir magnets whose only connection to the location is that they are in the shape of an iconic local site, and dining options that are increasingly found globally. It is perfectly logical in this tourism environment that the Bubba Gump Shrimp Co., a restaurant based on a Hollywood film, sits atop the Peak Tower in Hong Kong and that multinational brands are found in the signature shops. Tourism is big business. Yet with spreading wealth and global connectedness, it is not just tourists who are moving, but attractions as well. Hong Kong is not just the site of an American chain restaurant, but a Disney park franchise as well, thus continuing the use of well-recognized, global signs (Figure 3.4).

Figure 3.4 Global Signs in Malmö, Sweden (by Nicholas Wise)

Shaping Tourist Perceptions

In spite of, or perhaps because of, the fact that we live in a world saturated with imagery, it is important to consider the ways in which spaces become places before we even see them in person. This is especially the case within the context of tourism, which trades on the notion of locales being special, unique, and "must see." From the Grand Tour to *1000 Places to See Before You Die* (Schultz, 2003), the distant worlds "out there" have always evoked travelers' desires. The rise of mass tourism and the companies that fostered this rise drew heavily on place-making/place-marketing. Paradoxically, tourism promotion, while emphasizing the idea of the unique, also falls into a reliance upon genres. Within the context of tourism places there is the production side and the consumption side, and semiotics and landscape theory can help us understand both. In a sense, tourism imagery helps to condition tourists' interpretations and behavior in place, creating what Jenkins (2003) refers to as the circle of representation.

Tourists who have *not* been preconditioned to understand how to interpret tourist sites are as rare as Panda babies in the wild. While the tourism market expanded in the late 19th century, the idea of sightseeing as opposed to traveling for a purpose can be traced much further back. As early as the 12th century, there was an understanding that religious pilgrims were also interested in seeing the sites (Bar and Cohen-Hattab, 2003). A Latin guidebook published around 1143 for pilgrims coming to Rome included not just the Christian churches and places where early Christians were martyred, but also highlighted the interesting pagan sites of Republican and Imperial Rome (Morgan and Gardiner, 1986). The itinerary of the Grand Tour was well established among the aristocrats and petty aristocrats in Europe and was reinforced by images in the paintings, travel logs, and novels, that travelers brought back (Chard and Langdon, 1996; and Ousby, 1990). From their earliest inceptions, tourism industries have been shaping tourist interpretations of potential destinations by establishing repeated and circulated signs of place.

With the development of mass tourism in the late 19th century (Van Aalst and Boogaarts, 2002), the guide and guidebook were means of initiating the middle class—and sometimes working class—tourist into the proper way to tour and into what ought to be seen (Koshar, 1998; and Travlou, 2002). The segmentation of guidebooks began early, a distinction that has continued throughout the course of the 20th and early 21st centuries such that people often become very committed to a series because they have the sense that the publisher understands their particular view of what it means to be a tourist (Ashenburg, 2001; Stewart, 2005; and Jack and Phipps, 2003). The *Blue Guides'* appeal is to the pedantic among us who want to know architectural details, and cultural history that is more fleshed out than the brief snapshots given in other books, while those who like pictorial guides have flocked to the recent proliferation of the *Eyewitness Guides*. Parsons, in analyzing guidebooks for the city of Budapest, (2005, p. 99) wrote that,

> ... the Eyewitness series (British with translations into major European languages) most obviously moved the genre forward in terms of production

> values, with its excessively semiotic [sic] approach to communicating information and its sophisticated axonometric plans of major buildings. Virtually no latitude is allowed to its writers, whose contributions often read like sound bites to accompany the illustrations.

And those who want to avoid "the golden hordes" and find more "authentic" scenes of life in a place rely on *Lonely Planet* (Ashenburg, 2001), though they often find themselves off the beaten path and in the company of like-minded fellow travelers. The rise of the Internet and sites such as TripAdvisor and Virtual Tourist, with their seemingly unbiased crowd-sourced recommendations about the best places to stay and eat, may signal a break in the dominance of the curated guidebook, and as such, illustrates not just personal accumulated information about a site, but a means of socializing the collateral information process and illustrating its intersubjective nature.

The common denominator in tourism guides, whether of the analogue or online version, is the assumption of a set of expectations about both the particular locale and the ways in which one interacts with that locale when one is on tour, as opposed to when one is on business or is a resident. As a tourist, one should see the sites, sample local specialties, take pictures, and buy souvenirs. This is what the guidebooks specialize in. The 5th Edition of *The Rough Guide to Poland* (2002, n.p.) promises "up to the minute accounts of all the sights, from the fast-changing cities of Warsaw and Kraków to the laid-back lakeside resorts of Mazuria" and "critical reviews of restaurants, bars, and accommodation in every price range." *Frommer's 17th Edition to Scandinavia* (1997, n.p.) offers the following list,

> Everything You Need for the Perfect Trip:
>
> - Accommodations for every taste and budget: old world palace hotels, historic woodland inns, lakeside resorts, and family farmhouses
> - The best in authentic dining, from hearty country breakfasts to traditional smorgasbords
> - A complete sightseeing guide, from Lapland villages to Hamlet's castle – with art museums, classic market towns, Renaissance castles, and Viking ruins
> - The best of the majestic outdoors: spectacular drives, fjord cruises, reindeer safaris, Finnish saunas, and more
> - Shopping at the source for Royal Copenhagen porcelain, Norwegian knitwear, and Swedish glass.

Even the 7th Edition of the *Blue Guide to England* from 1965, tells readers

> … In this the seventh edition, the traditional arrangement in routes based on the road system, the sheet anchor of the Blue Guides series, has been preserved … the town plans have been produced as usual in co-operation with Bartholomews.

> Practical information about places to stay and to eat, bus connections, prices of admission, and dates of special events are included; a bibliography suggests useful reading; and a comprehensive index of places and persons is provided. The *Blue Guide to England* not only serves the hurried visitor who intends to 'do' the whole country in a month, but will be a constant companion to the regular traveler up and down the land. (n.p.)

Clearly the "hurried" visitor who only has a month to speed through England is no longer the norm, and instead guidebooks offer quick versions of sites, with suggested itineraries of the "must sees" so that the tourist can return having seen what one should have seen, and having taken the expected photos. It is in this vein that Jenkins (2003) speaks about the circle of representation as a process by which tourists actively participate in the reproduction and circulation of tourism iconography. The need to picture oneself on site is part of the modern tourism act, a practice we also see interrogated in Chapter 8 on touring Florence. The recent appearance of a fake skyline in Hong Kong for tourists to take pictures in front of, when the actual skyline is hidden behind a cloud of smog (Schiavenza, 2013), underscores the importance not of "being there" but of producing proof. This example leads us to think about the ways in which tourism is not only an embodied experience of place but one that is connected to a larger set of practices that signify being a tourist.

Conclusion

We are all, to some extent, semioticians. As sentient beings we cannot help but interpret the natural and built environment around us. We are in a constant state of semiosis, as Peirce defined it, making meaning from the world. Much semiosis goes on in the background, almost without our awareness, but at times we do bring this process to the fore. Being on tour is one of those times, as we are physically dislocated from our daily lives. But we bring with us a whole cache of information and expectations of what it means to be and perform as a tourist, some shaped by guidebooks, others informed by previous experiences and communication with others. On the ground, however, as a tourist one is confronted with material realities and sets of circumstances that require (re)interpretation pushing the boundaries of one's collateral information and thereby adding to it. Yet, we are not alone in this endeavor. In most instances we are accompanied by others—tourists, locals, and/or guides—as well as guidebooks and technologies that may assist in our interpretative efforts. As such, semiosis is also intersubjective. Further, as Peirce reminds us, semiosis is never complete. The meaning-making process persists upon returning home in the re-telling of stories and continued regard for souvenirs and photographs.

We suggest, as do other tourism scholars (see MacCannell, 1976; 1999; and Culler, 1981), that Peircean semiotics is, therefore, a more appropriate framework for approaching tourist meaning-making processes. Peirce's semiotic extends beyond cognitive associations to actions, or habits, that result from interpretation. Further, he was particularly interested in the ways in which interpretations are never complete but are dynamical. If we can be allowed to stretch the concept of genre, we might argue that tourism acts in the same manner, setting up criteria for creating and assessing places. A tourism site is successful if it fulfills the expectations that the tourist brings along. In this way, the performance of tourism impacts tourists' interpretations of places by accounting for the selective interests that they have for seeking out experiences.

As such, Peircean semiotics adheres well to a landscape perspective. It accounts for meaning-making processes at various scales—personal to collective—as well as how meaning changes over time. From the previous chapter, which attended to the theoretical roots of ritual aspects of tourism, we now have evidence for the ways in which rituals are communicative, and further, for the importance of considering the performative aspects of ritual in regards to the production of space. In the following chapter we will focus on the ideological aspects of the social construction and marketing of tourism sites. As such, semiotics resides clearly in the background of this analysis, informing the questions of why sites are chosen as tourism attractions and the processes through which tourist preconceptions are informed via marketing campaigns and onsite, interpretative materials.

Chapter 4
Ideology and Tourism Sites

Introduction

Theorizations of ideology have shifted and evolved tremendously over the last 150 years. Marx asserted that ideology is "false consciousness," a selective view of reality laden with the ideals and values that reproduce class relations. Investigations of ideology, however, were not his primary project. Nevertheless, subsequent scholars have built upon his notions of the concept, thereby establishing various theories of ideology. Gramsci (1971), for example, developed a concept of hegemony based on the ways in which ideology is negotiated. Through processes of consent and coercion, rather than domination, ideology surfaces through "sites of struggle." As such, Gramsci illuminates the social processes of ideology in terms of subordination and contestation, as well as the contradictions between ideology and lived experience. Like Gramsci, Althusser rejected Marx's economistic ideology. Examining the ways in which nation-states function, Althusser posits that states rule through apparatuses—ideological and repressive. Unlike Gramsci, Althusser emphasizes the processes of interpellation, or rather the habituation of subjects. Later contesting the work of Althusser as deterministic, Lefebvre argues that ideology is (re)produced through social practice, and as such, social practice produces space, making it more susceptible to change. This chapter works through the concept of ideology using concepts theorized by a number of scholars to consider ideology's implications for tourism.

The question as to why some places are tourism sites and others are not is a complex one involving issues of ideology, particularly as ideology relates to identity, mythology, and signs. We use the last term, sign, in the Peircean sense as something, anything, that stands for something else (see Chapter 3). To frame our argument, we draw upon Victor Turner's (1969) classic definition of a performance as an event in which actors communicate via signs to an audience that might or might not understand those signs. Performances are communicative precisely because the audience and the actors share a common understanding of the signs used. In tourism, however, the actors and audience are not so easily delineated. Tourists are not simply an audience, sitting and watching performances held before them; they enact tourism places, they perform tourism sites as signs (Lau, 2011; and Knudsen and Rickly-Boyd, 2012). These tourism performances are not spontaneous but informed by promotional devices, place circumstances, and individual motivations. Despite the differing contexts of tourism from Turner's anthropological investigations of indigenous rituals, signs remain central to tourism performances of place. However, the ideology of sign production and communication in tourism becomes more difficult to tease out. The previous chapter focused on the dynamics of tourist meaning-making through the interpretation of sites as signs. Here we aim to investigate another side of this relationship—the role of ideology in the constitution of tourism sites as signs. We begin by considering why and how some places become tourism sites. The institutional choosing of tourism sites is set within a number of ideological frameworks, notably place identity and myth. This selection process leads to questions of the mechanisms by which tourism places are marketed and the ideological devices employed, specifically tourism brochures. Finally, we make note of tourist agency in this scheme. While a focus on ideology can lead to rather deterministic conclusions, it is worthwhile to consider the choices that are made by tourists in response to individual motivations, desires, and limitations.

Ideology and Place Identity

In the realm of tourism, the construction of place identity is central. At the heart of identity politics is the distinction between "self" and "other." In the construction of place identity, some sites are chosen and projected to potential tourism audiences, whereas others are ignored and even hidden from view. The various tourist boards and quasi-governmental committees that act to determine which sites are "must see" are seldom materially visible to the touring public beyond their stamps of approval on promotional devices. Nevertheless, these organizations exist and act. In the absence of their work, there would be no "official" guides to the cities awaiting tourists disembarking the air flights and cruise liners, no city slogans encouraging a particular tourist mindset and experience, no advertisements of tourism locales, and no brochures explaining the significance of individual sites. Tourism boards and committees thus produce a set of signs for tourism performance at any given locale. This process, however, immediately raises a dilemma, since a successful performance requires signs to be understood in the same way by both actors (the tourism boards) and audiences (the tourists). Most generally, tourism traffic is categorized as either domestic or international. Tourism sites signify very different relations of identity to a domestic audience as opposed to an international one. When tourists hail from the same country as tourism boards, as is the case with domestic tourism, then the glue that binds actors, signs, and audiences together is a set of ideological apparatuses based upon national identity. In the case of international tourism, there may be considerable slippage between the tourism site and tourist, between sign and audience. We explore that slippage in greater detail below.

The relations of identity politics and tourist origination are exemplified in Light's (2000; 2001; 2012) work on the role of tourism in post-Socialist Romania. Rebuilding Romania following its Socialist period, tourism has figured prominently, Light argues, both in changing the country's image on the international stage as well as internally recontextualizing its history and projecting the future to the body politic. This interrelationship has posed a particularly difficult challenge, considering the iconic images of place identity with which tourism boards must work. From the stereotypical, pop culture representations of Dracula to the *Casa Poporului* (House of the People), a remnant of totalitarian rule that is among the largest buildings in the world and the location of populist riots that overthrew Ceausescu's government, these sites cannot be ignored. Their associations can be changed, however. The "House of the People," as a colossal structure, is used to highlight the craftsmanship and architectural talents of Romanians to both internal and external tourism audiences. However, as the primary site of political revolution, it also symbolizes the power of the Romanian people, fortifying an internally projected national identity and recontextualizing a period of time many Romanians would rather forget. As a revolutionary site, it symbolizes to an external audience more simply the power of democracy and thus connects Romania politically and economically to Central Europe.

The role of the Dracula image, however, is a greater challenge to Romania, according to Light (2012). While Dracula is a veritable force in drawing international tourists, it also conjures images of vampires and the supernatural, thereby contrasting efforts to project a modern, developed, cosmopolitan Romania. Thus, Light's research in Romania illustrates the layers of politics that drive the projection of place identity in tourism. Such signs of national identity signify differing politics to domestic and international tourists. Despite political and cultural pasts that some would rather see forgotten or even erased, such powerful signs of place identity, once in circulation, cannot be ignored. Tourism organizations, however, can work to alter the signs' meanings by associating them with new sets of signs. In so doing, tourism organizations enlist the ideological power of signification. To better understand the relations that underlie identity politics and tourism marketing, we suggest greater attention to the mechanisms that (re)produce and circulate ideology.

Ideology, Ritual, and Tourism

Discussions of the linkages between ideology and tourism are not unknown in the tourism literature. Important contributions have been made by, among others: Pritchard and Morgan (2001), Shaffer (2001), Light (2001), Ateljevic and Doorne (2002), Pretes (2003), Palmer (2005), and Jokela and Linkola (2013). Ideology functions by habituating individuals as citizen/subjects into society through a number of overt and nuanced rituals of daily life. Althusser (2001 [1971], p. 19) makes the distinction between repressive state apparatuses that work through violence, such as the police or military, and those that work through persuasion—ideological state apparatuses—including family, educational institutions, religious organizations, and the media. He observes that nation-states generally rely upon ideological state apparatuses for their rule, employing repressive apparatuses as a last resort (or at least purporting to employ them as a last resort).

Ideological state apparatuses, and the processes of signification they enable, work through ritual. Ideology, Althusser argues, is made manifest in practices and rituals, and it is only in these practices and rituals that ideology can exist (Althusser, 2001). Because the power of ideology rests on its being habitual, Althusser claims it is rarely recognized by those who are its subjects: "ideology never says, 'I am ideological'" (Althusser, 2001, p. 49). He terms the habituation of individuals into society by the ideological state apparatuses "interpellation." Once individuals are interpellated, ideologies are represented (and recalled by those interpellated) through their signs, a process Althusser terms "hailing." In this way, ideology becomes ritualized—the interpellated recognize that ideology hails them precisely because they already have assimilated the ideology (they are interpellated) and already have come to recognize ideology's signs, its hailing of them. This theorization is applicable to tourism as a ritual, as tourism organizations choose signs of place identity, which hail tourism audiences to various degrees.

In particular, this chapter suggests a considerable difference in the ability of tourism sites as signs to hail domestic versus international tourists. In the case of domestic tourism, tourism sites function as signs that hail both actors (tourism boards) and audiences (the domestic tourists). This process complicates MacCannell's theory of sightseeing as a modern ritual (see Chapter 2) in productive ways. Participation in the collective ritual of visiting the same significant sites, he argues, offers the potential for social integration to those alienated in their everyday lives. MacCannell's ritual considers social integration at the broadest of social scales; yet, to consider the differences between even domestic versus international tourists suggests varying levels of potential social integration, as is evidenced in the case of tourism in Romania (Light, 2000; 2001; 2012). This difference is most salient in cultural and heritage tourism, as these types of tourism are driven by identity politics. Heritage tourism is motivated by one's desire to experience a greater sense of collective identity through touring sites of historical significance (see Poria, Butler and Airey, 2004; 2006), whereas cultural tourism is quite the opposite—tourism to see and experience an "other" (see Bruner, 2001). The same sites of tourism can function in disparate ways depending upon the ideological rituals performed. Historic locations in China, for example, can be both heritage sites for Chinese tourists and cultural tourism sites for U.S. tourists. In either case the same site is ideological, yet its ritual performance by tourists offers different scales of social integration.

Such an examination of tourism sites as signs based solely in Althusserian notions of ideology and ritual, however, does not address the ways in which the significance of sites as signs change over time. Indeed, as Light illustrates in Romanian tourism promotion, the relationship of tourism sites to national identity changes with political and social circumstances, yet institutions can also manipulate their ideological associations. Incorporating Barthes' concept of myth along with Žižek's notions of Master Signifier and Sublime Object allow us to further investigate the nuances of ideology and tourism sites.

Place Identity and Myth

In developing a basic theory of tourism sites as signs, MacCannell (1976) utilized a semiotic framework. Similar to a sign, which represents something to someone, MacCannell argues that tourism attractions mark sights to tourists. We agree with such an analysis, and the previous chapter further develops a Peircean semiotic in regards to the tourist experience. Working through the mechanisms that produce tourism sites as signs, however, we suggest it is useful to consider how signs build upon one another referentially to function as myth.

The social construction of tourism sites always involves the mobilization of myth, argues Rojek (1997). According to Barthes (1972), myth is a form of speech. It depoliticizes sets of relations and makes them appear natural. Myth works through hierarchical ordering so that a sign becomes a signifier of a higher order sign. This ordering can, and indeed does, continue over time as signs take on new referents. Additionally, Barthes contends, one cannot deconstruct myth so as to break its communicative power. Only by further building upon myth can its meaning be altered. It is in this way that "…myth has the task of giving an historical intention a natural justification, and making contingency appear eternal" (Barthes, 1972, p. 142). Thus, in the case of tourism, sites are signs that communicate through myth. Tourism sites are not a single sight, but they are landscapes composed of a series of objects—the monumental and the mundane working together referentially. Because Barthes' myth is based in Saussurian semiology, it tells us how tourism sites as signs communicate ideology. It does not, however, offer a means to deconstruct the mechanisms of the mythology of tourism places as both material and imaginary landscapes. We thus turn to the work of Žižek as way to tease out the material/imaginary relations of tourism places.

From the work of Žižek (1989), we argue, tourism sites are often Master Signifiers that make reference to one or more Sublime Objects. For our purposes, a Master Signifier is a physical object that stands for some fundamental national myth or imagined ideal, the corresponding Sublime Object. A Sublime Object is a thing or person that is beyond reproach or contestation—its signifying power is beyond comprehension. For example, the U.S. flag stands for the United States and all its rituals and practices, its history, and so on. The Lincoln Memorial stands for Abraham Lincoln, but also for the American Civil War. And, of course the Capitol, the Lincoln Memorial, the White House, and the Jefferson Memorial make a cross, or if you prefer an X, as in "this is the place"; this is the central axis of the United States (Figure 4.1). A tourist standing at that X is in the presence of the essence of what the U.S. stands for and has stood for—its Constitution, its Bill of Rights, its delicate balance of power between the executive, legislative, and judicial branches, and its striving for a more perfect union from a population of enormous diversity through a continuous process of emancipation. Importantly, this essence is a privileged one, manufactured through ideological state apparatuses, that continues to serve pernicious racism and inequality. Myth is thus the seemingly "apolitical" glue that binds Sublime Objects and Master Signifiers, imbuing those Master Signifiers with

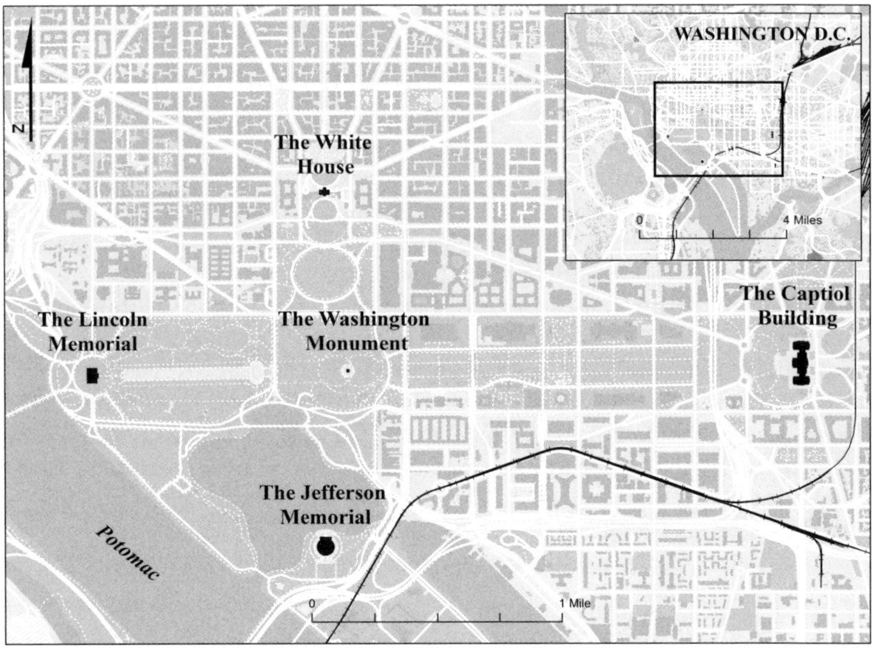

Figure 4.1 National Mall, Washington, D.C. (by David P. Massey)

the mythological properties of the Sublime Object and binding both into the corpus of ideology. It is what connects the tourism sites we visit with the reasons that site has been delimited as a place of importance. Recognition of the X's ideological investments may come more easily to a domestic tourist than an international tourist.

While notions of Master Signifiers and Sublime Objects account for the ways in which tourism sites as signs communicate through the relations of myth to material objects, there are also more subtle communications of place identity in the landscape of tourism sites. Additionally, ideology functions at the aesthetic level of practice. Tedman (1999) contends that a strictly Althusserian analysis of ideology gives too much potency to the *ideas* of ideology and not enough to the *feeling* subject. We acknowledge aesthetics as an important component of emotion, and as such, it is aesthetics to which we now turn.

Aesthetics

Historically considered the science of perceptions, aesthetics is an area of philosophy that is concerned with taste and the nature of art (Brady, 2003). An aesthetic level of practice speaks to sensory awareness, feelings, and emotional responses. Tedman (1999, p. 62) suggests aestheticized ideology is more stable, as "feelings are far harder and far slower to change than ideas." As a result, "feelings are at stake in a battle that is fought out in subtle sensual maneuvers" towards an affective ideology. The aesthetic level of practice is particularly salient in regards to tourism. As tourists, we leave "home" and experience another place, in which we attempt to make sense of a series of signs, some marked by tourism organizations and some not, working as an ensemble to create place. Such a mix of signs is important to validating a sense of place, from subtle, mundane signs of everyday life to grand, overt symbols of power (Rickly-Boyd and Metro-Roland, 2010). More importantly, many of the landscape features of tourism places were not produced for this industry but relate, instead, to local or national societal relations, and have only more recently been employed in projections of place identity. That is, much of the architecture and landscape of tourism places is designed to reify internal social relations. Interestingly, the same aesthetic modes used internally also convey particular emotional and behavioral responses in tourists. Thus, in considering the role of aesthetics in tourism, we wish to focus on just three of the broadest categories of sites—the beautiful, the pastoral, and the sublime.

The beautiful serves as the first category of sites—it is created by signs that communicate the relationship of a society as a political entity to its deity. As such, "beautiful" sites function as creation stories for the society and serve to justify its respective form of rule of law. Such tourism sites are as often as not part of the built environment, and as such, they have an intentional design process that is relatively easy to decode. Returning to the example of Washington, D.C., consider the U.S. Capitol Building.

Figure 4.2 The U.S. Capitol Building (by Lauren Clark)

It has all the markings of the "beautiful" aesthetic. Note the dome and spire reaching to the heavens and the balanced symmetry on either side of the dome, signifying the equal status of the Senate and House of Representatives. The neo-Classical design of the building is also noteworthy in that it makes reference to Greek democracy through use of the flat roofs over the legislative chambers and Ionic columns.

If the beautiful is an aesthetic about the relationship between deity, power, and humankind, then there are other aesthetics that signal other things. Consider, for example, this scene from rural England.

Figure 4.3 The Pastoral Aesthetic, Southern England

Here, a select group of insiders to the scene have, over time, built a pastoral cultural landscape. This relationship of society to the pastoral is yet another important aspect of ideology that hails through aesthetic principles, and defines the historical basis for the various rituals and practices of the society. Mid-summer fields bordered by dense hedgerows generate nostalgia for a largely imaginary past when life was simpler and the pace was slower, which hides the reality of modern industrial agriculture. A footpath and distant homes secure a human presence in this landscape, while Norman-era ruins hint to a rich history of this place.

Whereas the pastoral ties humanity to its past and evokes nostalgia, the sublime relates humans and nature. Nature involves great power, incomprehensible size and number, and the unspeakably strange and unknown. Nature is both sacred and dangerous, both ennobling and democratic. As part of the natural landscape, too powerful to be harnessed or shaped by the state, sublime sites become part of the national landscape, examples of nature "owned" but not under control. These include vast wilderness, trackless desert, deep caverns, thundering waterfalls, temptuous seas, and destructive volcanoes.

Figure 4.4 The Sublime Aesthetic, Yellowstone National Park (by Isaac Rooks)

In addition to its connection with nature, the sublime is often used in conjunction with the power of a deity or, once removed from deity, the State. Additionally, the incomprehensibility of the sublime may be associated with the vastness of large nation-states (for example, Russia) which cannot reasonably be visited in their entirety by all their inhabitants in a lifetime, of the vastness of the population of a single nation of peoples (for example, the Chinese) who cannot reasonably know one another outside of some imagined community (Anderson, 2006).

Each of these aesthetics—the beautiful, the pastoral, and the sublime—utilizes a particular set of signs. While more generally employed in landscape and architecture to communicate social relations to the body politic, such sites are now often redirected in tourism marketing. As such, each is a toolbox for use by those who wish, by the act of selection as a tourism site, to communicate a relationship within society to an audience. Indeed, national imagery in the United States, for example, combines all three aesthetics, with more emphasis on the sublime than France or Spain, suggesting the greater significance of the frontier to national identity. Thus, it is in the marketing of tourism sites that the role of aesthetics functions in the abstract, drawing on associations tourists may have with particular signs of place. With regard to tourism marketing, the following section focuses on the devices of tourism promotion.

Tourism Marketing and Promotional Devices

Marketing, advertising, and public relations have long been tied to notions of ideology. As Chambers (2000) notes, "tourism is not only [about] economic gain, it is also about the uses of power and about the ways people choose to represent themselves" (p. 31). While "branding" may seem a concern of corporations alone, nations and cities, too, approach similar issues as they choose representative sites to promote in the tourism industry (Hall, 1998; 2002; Olins, 2002; Beeton, 2004; and Konecnik and Go, 2008). Hall's (2002) study of former Yugoslavian countries' tourism development efforts, for example, highlights the difficulties of "rebranding" national images after considerable political change and conflict (see also Light, 2000; 2001). As such, tourism branding is inherently ideological as tourism organizations choose among multitudes of potential signs of place, highlighting some and ignoring others, in an effort to communicate a clear and concise message of place identity to tourism audiences. While some destinations work to further refine their brand through images in brochures and on travel websites, others, such as the former Yugoslavian countries (Hall, 2002), work to re-define a place through new image associations. The success of tourism sites rests on the projection of myth, and, asserts Hughes (1992, p. 34), "lies in its ability to link up with a store of perceptions and experiences that are already embedded in commonsense understandings of geography."

Lew (1991, p. 126) suggests that if the essential elements—tourist, site, and marker—are present, "virtually anything can become a tourist attraction." Tourism marketing, however, is not as simple of a task as crafting a place identity, choosing a set of landscape features, and compiling a brochure or website. The ideology of tourism sites should be given greater attention in this process (see also Pritchard and Morgan, 2001; and Ateljevic and Doorne, 2002). To begin, most tourism sites aim to attract a variety of tourists and construct a number of place identities as opposed to a single identity. A multiplicity of meanings is thus marketed through a series of promotional devices. Pritchard and Morgan (2001) illustrate just such a process in their examination of tourism marketing by the Wales Tourist Board. By comparing branding representations presented to two different audiences—American and British—the authors find two discernible place identities of Wales, which reflect differing power relations. While the promotional devices aimed at American tourists emphasize Welsh national symbols (the daffodil, the leek, the Welsh dragon, and Welsh flag), along with Celtic typeface, the Welsh language, and heritage (myths, legends, castles, and musical traditions), the promotional materials for British audiences use these representations sparingly, highlighting instead the area's scenery, people, and rural life. Pritchard and Morgan argue that these differing representations of place based on intended audiences affirm the power relations in tourism marketing that relate to complexities of historical relations and power dynamics. In this case, the colonial relationship of Wales and the United Kingdom has a history of subjugation, prejudice, and violence through state power, and from which British media continue to portray the Welsh as "other." In the construction of place identities for tourism, the authors suggest the ways in which Welsh tourism organizations work within the ideological constraints of Welsh-British relations to downplay social difference and emphasize scenery and hospitality. It is, then, in the representations of place identity intended for American tourists that the Wales Tourist Board discursively constructs a sense of place identity based on national heritage and difference within the United Kingdom. Tourism marketing strongly evidences the close relationship of ideology and semiotics.

Semiotics and Tourism Marketing

MacCannell (1976) first proposed that the interaction between tourists and tourism sites is an exercise in semiosis. Building from the previous chapter that focuses on tourists and the processes of semiosis, here we are interested in the ways in which tourism organizations play with semiotics in marketing campaigns and promotional devices. The primary role of promotional materials is to induce a desire to visit a place, which is accomplished by engaging a set of aesthetics and sign associations in potential tourism audiences. A Peircean semiotic perspective suggests this intention to provoke interest and curiosity is connected to recognition; that, in the case of tourism promotion, the potential audience must be able to interpret the images (*objects*), at least in the most general of ways. However, the *interpretant* is not a "thing," but rather a feeling. Thus, we can return to the discussion of aesthetics. Most of the signs that comprise promotional materials are cognized within the realm of aesthetics—the beautiful, the pastoral, and the sublime. Indeed, browsing promotional materials, we tend to find imagery of cities (the beautiful), the countryside (the pastoral), and wilderness (the sublime), or an interplay of scenes of homogenous modernity (the beautiful) with the exotic (the sublime), and traditional (the pastoral). Given the individual motivations of tourists at a particular place, different aesthetic registers will better engage their interests. As a result of the diverse tourist market, tourism organizations design a variety of promotional materials. Indeed, Jenkins (2003) examines the similarities and differences among Australian tourism brochures, in particular those aimed at the mass tourism industry and those more specifically targeting backpackers.

Further, commercial images inform particular tourist gazes (Urry and Larsen, 2011). The concept of the "tourist gaze" stems from the Foucauldian notion of the medical gaze, suggesting that "looking is a learned ability and that the pure and innocent eye is a myth" (2011, p. 1). In other words, "People gaze upon the world through a particular filter of ideas, skills, desires and expectations. Gazing is a performance that orders, shapes and classifies, rather than reflects the world"

(2011, p. 2). As a result, commercial images used in promotional materials suggest particular ways of seeing by tourists. Urry and Larsen (2011) note that such images work two-fold in that they produce the desire for travel but also script and frame potential tourism destinations through imaginative geographies. That is, they produce place-myths. The seductive power of place-myths, however, lies in the ability of the images presented in promotional materials to appeal to individual tastes. The tourist gaze is not unidirectional. Commercial images must first grab the attention of potential tourists; as such various gazes attract different tourists at different moments. This lends further support to the development of multiple promotional campaigns by the same destinations, each presenting a distinct gaze: romantic, collective, anthropological, environmental, family, and so forth.

While promotional devices are designed to prompt interest, and in so doing employ aesthetic modes, on-site materials operate through another set of ideological mechanisms. On-site materials are intended for use at a particular attraction, for visitors who are already there, on the ground. They do not promote the site as a potential attraction, but work to assist in tourist interpretation by establishing historical context and cultural significance. The relationship between promotional devices and on-site materials is taken up in Chapter 7 with the examination of a Danish tourism promotional video and various multilingual on-site brochures gathered at number of attractions.

On-Site Materials

On-site materials are distinct from promotional materials in that their primary function is to educate, rather than entice. Once tourists have arrived at a particular attraction, interpretative materials are employed to aid tourists in making sense of the site. Thus, while promotional materials work abstractly, through sign associations, on-site materials are more concrete, providing context for interpretation.

Attempts at framing tourists' interpretations, however, are inherently ideological. By providing further detailed information for tourists on the ground, at particular sites, tourism organizations foster a *sensus communis*, a shared experience or meaning (Sharpe, 2005). While more detailed interpretative materials may appear to offer greater transparency in relating the significance of a site, those materials moreover illustrate the fashioning of a context by what is included and what is excluded. As such, efforts to encourage similar interpretations, to encourage a *sensus communis* are met with highly individualized reactions. Chronis' (2005; 2008; 2012) work at Gettysburg illustrates these dynamics. Despite managerial efforts to streamline the narrative of Gettysburg as a pivotal site in the American Civil War, the author observes tourists do not simply adopt the story presented to them but contest and even reject some elements. Chronis, thus, terms this process a co-construction of the Gettysburg storyscape. Interpretation of the battlefield, including narration by certified guides, is intended to contextualize the events that took place. Yet, tourists arrive at Gettysburg with a range of knowledge, from public education to popular history documentaries and non-

Figure 4.5 On-Site Interpretative Materials, Spring Mill Pioneer Village, Indiana (U.S.)

fiction books, as well as personal interests in history. On-site interpretations that support individual understandings are often accepted, whereas interpretations that refute understanding are more likely to be scrutinized and even rejected. On-site materials may offer an authoritative voice, but they do not have the power to determine individual interpretations.

A Note on Tourist Agency

Discussions of ideology wax deterministic. Althusser is no exception, as he is noted for deeply structural theories of the subject. In the discussion of ideology and tourism, it is worthwhile to consider the agency of the tourist in this scheme. From the previous chapter on tourist experience, which focuses on semiotics in meaning-making, to the following chapter which aims to bring together the tourism site and tourist experience as an enactment of place, we have taken care to speak not to ideological determinism but to power relations. Such relations form complicated webs of consent and resistance, nuancing the ways in which signs are managed by governing bodies.

While continuing a similar theoretical trajectory as Althusser, French Marxist Henri Lefebvre offers a productive transition from Althusser's ideology and interpellation to the social production of ideology. Lefebvre's work attends to a greater sense of agency while still supporting the ideological assessments of tourism sites made throughout this chapter. Lefebvre's (1991) *The Production of Space* suggests space is not an absolute given but produced through social practice. Thus, he argues, social space cannot be defined in terms of the projection of ideology. While ideology may direct the locations for particular activities, "ideologies do not produce space: rather, they are in space, and of it. It is the forces of production and the relations of production that produce social space" (p. 210). Lefebvre is highlighting the importance of the performance of ideology. Ideology does not do things, but suggests and encourages particular behaviors. Indeed, as MacCannell (2001, p. 24) asserts, "What we gaze upon as tourists may have been arranged for us in advance; we may go there precisely because other tourists have gone before us; but we remain free to look the other way, or not to look at all. And we can disrupt the order of things." We can perform as per the norm (naturalized behavior), we can contest, adapt, and change behaviors, thereby pushing ideological boundaries, or we can protest and act in opposition to what is expected. That is not to say that our actions are not informed by other ideologies and/or power relations, but to emphasize that it is through our actions, our agency to act, that social space is produced, and as such, is continually changing.

Thus, tourist agency in relation to the ideological framing of tourism sites can be witnessed in the social and spatial dynamics of staging, a conversation further elaborated in the next chapter. Tourists strive to peak behind curtains and explore the inner workings of place. As tourists, we know that a scene hides as much as it reveals, that looks deceive. Thus, we resist the limitations of tourism sites, in their spatial boundaries, social staging, and ideological framing. This resistance, MacCannell (2001, p. 31) argues, is strongest in "the desire to get beyond touristic representations," and is particularly evidenced in tourist photographic practices, in which tourists photograph the iconic but also the mundane, the extraordinary and the ordinary.

We may also return to the discussion of Chronis' (2005; 2008; 2012) work that considers the co-construction of Gettysburg. While the narrative presented to tourists is an "official" interpretation of historical events and context, tourists react individually, some accepting the narrative in this form, others rejecting it. Similarly, Rickly-Boyd (2014) finds that managers of Spring Mill Pioneer Village have established a particular history that is to be communicated to tourists; yet, the interpreters who work in the village stray from this narrative according to the tourists they confront. Indeed, interpreters adjust the village's story based upon what they think individual tourists want to hear more or less about.

In considering tourist agency in relation to ideology and tourism sites, we take a step further and turn to Foucault and his rejection of ideology. While we do not reject ideology generally, Foucault's arguments do move conversations about the ideologies of tourism forward in productive ways. Foucault's contention with ideology, in the Marxist sense, is three-fold: it presupposes a truth behind "false consciousness"; it refers to "something of the order of the subject"; and it "stands in a secondary position to something which functions as its base" (1994, p. 60).

In this critique of ideology, Foucault rejects universal notions of truth. Domestic tourists may not, for example, have more intimate relationships with their nation's history than outsiders do.

In order to mitigate Foucault's contentions with ideology and the Marxist perspective presented in this chapter we suggest the following observations of the relationship of post-tourists to tourism sites. In research on post-tourists (see Haldrup and Larsen, 2003; Jenkins, 2003; Larsen, 2005; and Rojek and Urry, 1997), scholars have found that by capturing the same iconic images used to advertise tourism destinations, yet playing with photographic practices in terms of staging and posing, these tourists acknowledge the power relations of tourism media. There is a recognition of the ideologies (rather than ideology, singular) that frame tourism destinations, marketing, and behavior, which suggests a more creative, subjective, and empowered engagement with the semiotics and ideology of tourism (see Crouch, 2005).

Conclusion

Ideology is central to identity politics, and as such, should be given specific attention in tourism studies. From the construction of place identities to the marketing of tourism attractions, tourism is inherently ideological. Tourism organizations, both those visible and hidden from view, act by choosing specific sites for promotion. But in the promotion of tourism sites, place identities are neither singular nor stable; they are multiplicitous, framed for distinct audiences, and continually adapting. Indeed, the very fact that tourism marketing employs myth makes such a dynamic possible. Myth, argued Barthes, continually evolves, building and growing referentially. Thus, tourism sites function as signs. And, indeed, the previous chapter establishes such a relationship in tourists' efforts at meaning-making. Tourists make sense of tourism sites by drawing on a cache of previous experiences and knowledge. Yet, as this chapter has suggested, tourism sites are not simply empty symbols for tourists' projections of meaning, but are chosen and framed to encourage quite specific interpretations. While tourism marketing is aimed at eliciting curiosity and interest through engagements with potential tourists' associations with abstract signs, such as aesthetic modes of the beautiful, sublime, and pastoral, on-site promotional materials work to limit the possibilities of interpretation and meaning-making. It is the use of Master Signifiers which comprise the substance of promotional materials. But here a problem of interpretation intervenes in that all those who may serve as audiences to promotional materials may not make connections to the underlying Sublime Objects. That is, we all carry with us various sign associations, some differing only slightly, others more drastically. This clash with the interpretative framing provided makes room for tourist agency. While tourism sites are ideological, tourists must perform tourism sites. It is through tourist performances that ideology is made manifest, perpetuated, altered, contested, and rejected. In the following chapter, we turn our attention to the relationship of tourism performances to tourism places.

Chapter 5
Performing Tourism Places

Introduction

Performance theory has been used in a number of ways across many disciplines including geography, anthropology, and communication studies. "Performances call forth multiplicity," Leavy asserts, "and are open to multiple meanings, which are derived from the *experience of consumption*, which may involve a host of emotional and psychological responses" (2008, p. 344, emphasis original). So while Foucault (1994) suggested three distinct epistemes in his genealogy of knowledge—the Renaissance based on resemblance, the Classic period based on representation, and the Modern period based on structuralism—Yudice (2003) suggests a fourth episteme to characterize the global era—performativity. Thus, performance theories offer another way of understanding tourism, that is, by foregrounding the *doing* of tourism.

In this chapter, we aim to provide brief overviews of several ways performance theory—and its entailments—have been and could be applied to tourism. Like the broadly constructed "performances" Leavy describes, instances of tourism can be textualized, read, and analyzed in myriad ways. This chapter first considers tourism through the frame and metaphor of the stage—it then blends tourism with persuasion and activism, arguing that the tour is a productive site through which to interrogate specific structures that encourage or discourage agency. In the second half of the chapter, we discuss tourism as a performance of place and the performative turn in tourism studies; with and into this last critically important concept, we incorporate non-representational theory (NRT) and embodied practice. To conclude, we briefly summarize the relationship of performance to the previous theoretical chapters, namely ritual, semiotics, and ideology.

Tourism as Staged Performance

MacCannell's (1976) *The Tourist* was among the first to frame tourism as a performance. Inspired by Goffman's theory that social interaction contains front and back regions, MacCannell developed the metaphor of the stage to help explain sight marking and the problems of authenticity in modern tourism. His notion of staged authenticity presents six stages, from front regions where hosts and guests meet, such as hotel lobbies, to back regions where "work" is performed out of view of tourists, such as hotel kitchens. Although architecture supports these divisions, MacCannell contends that in many cases the distinctions between front and back have more to do with social roles and activities performed than architectural functionality. During most Birthright Israel tours (discussed in the next chapter), for example, groups visit Bedouin tents (Figure 5.1). The tents are designed largely for tourists, but the performances tourists encounter within them make the implicit argument that visitors are encountering an authentic lifespace.

In tourism, this dichotomy between front and back leads the tourist to desire to infiltrate the more "real" back spaces, resulting in the creation of touristic sites which offer varying degrees of purported backstage access. For example, zoos offer back regions where they purport to show animal care, but these are in actuality sets constructed solely for sightseers. While the viewing balcony at the New York Stock Exchange, on the other hand, is an example of a tourist setting, as it is a space marked off for sightseeing among an everyday activity. Both are, nonetheless, sets—"the only reason that need be given for visiting them is to see them" (MacCannell, 1999, p. 100). These sets, according to MacCannell, are hyperreal simulacra, "copies that are presented as disclosing more about the real thing than the real thing itself discloses" (1999, p. 102; see also Eco, 1986; and Baudrillard, 1994). In each of these examples, a zoo's staged back regions and the viewing balcony of the New York Stock Exchange, "work" is witnessed but not participated in; labors are observed but not directly engaged. Both the sets and individuals who create them are performative, but the performativity of the former can serve to obfuscate the labor of the latter.

Figure 5.1 A Bedouin Guide Hosts U.S. Tourists from Birthright Israel

Framing and Footing

Additionally and importantly, performance theory offers many ways to complicate the notion of the tour beyond the "staging" metaphor. While MacCannell was influenced by the work of Goffman in his analysis of tourism staging, returning to Goffman's theories offers tourism studies even more to consider with regard to the social framing of tourism encounters as well as the footing, or positionality, of individuals (tourists to locals) in these encounters. In *Frame Analysis* (1974) and *Forms of Talk* (1981), Goffman explores what can broadly be categorized as individual and group negotiations. Dealing both with the collective sociality of everyday life and more deliberate performative decisions of the individual, Goffman maps the concept of performance onto various banal interactions. Performances thus extend beyond the performer/stage/audience framework, generating very context-specific frameworks of their own. Social frames are defined as "guided doings" (1974, p. 22), and such frames are quite obviously guided on a tour. Goffman argues that frames not only delineate activities, they can transform activities. Frames designate full and partial participants in a scene, yet there is certainly no requirement that scenes take place on a physical stage. The performative frame as an organizational mechanism is therefore important to bear in mind in the study of tourism; as an endeavor, tourism often encompasses a delivery of information unfamiliar to the tourist (see Chapter 4). Thus frame analysis may be particularly useful for the study of tourism encounters, especially guided tours, as well as the ways individuals convey tourism experiences to others—that is, the way tourism is used as social capital in one's everyday life.

Noting that staged activity must be substitutable for unstaged activity, Goffman utilizes a variety of theater metaphors to argue that drama allows subordinate channels of communication to be staged. These subordinate channels include asides, collusions, and intentions—all rich and viable sites for tourism studies. Insofar as tourism is a generator of "flexible make-believe" (Goffman, 1981, p. 237), it replicates the framed activity of the stage—or in Goffman's paraphrased words, the main line of activity is carried out alongside labor (and presumably play) that happens outside of the frame. The "main line of activity" is the tourist's intended, semi-preplanned (or more "official") experience, whereas the "out of frame activity" encompasses all that occurs to make tourism so often appear to be so easy. While Kingsbury (2011) interprets part of the complex relationship between "tourists" and "workers" at Sandals' resorts as one of psychoanalytic *jouissance*, the frame analysis of performance theory presents a useful theoretical addition. Considering precisely who is within and who is without a given tourism frame is simultaneously spatial and political, carving a space for the non-representational theory discussed below.

In addition to frame analysis, Goffman (1981) provides a productive way to connect performance, ritual, and tourism through his notion of footing. He defines footing as a change in alignment that can simultaneously be a change in framing; footing encompasses the ability to decipher a situation or context and blend into or break oneself from it. By way of an example, Goffman uses President Nixon's imploring of journalist Helen Thomas to turn around and show off her pants suit before a public statement—he argues that in this case, women must always be prepared to change footing, or have their footing changed for them. In the case of tourism, footing as a concept is central to those forms of tourism that shift the positionality of the tourist. From "dark" tourism encounters with sites of violence and tragedy in which tourists are encouraged to imagine victimization and dehumanization to newer iterations of poverty or slum tourism that offer affluent tourists the opportunity to experience impoverished ways of life, tourists are constantly asked to shift their imaginations of themselves in the world.

As we move to consider the political dimensions of tourism—the interconnected politics of the tour, the tourist, the toured, and the frequently hidden labors that mutually sustain them—the language of footing provides a partial vocabulary for discussing the inter- and intra-cultural alignments of touring phenomena. As Silvio (2010), Gershon (2010), and others note, concerns about who authors and who animates texts are frequently—though often quietly—fueled by the politics of textual production. The study of tourism encourages consistent acts of entextualization: places themselves become texts, and publicity materials, souvenirs, and the like are read as such. Throughout this book, we nuance the interplay between texts, performance, and tourism, offering alternative readings of tourist places.

"Staging" as Metaphor

Building upon MacCannell's stage metaphor, some tourism scholars have worked to further integrate tourism, performance, and theater studies literature (see Crang, 1997; 1999; Desmond, 1999; Edensor, 2000; 2001; and Baerenholdt et al., 2004). In this vein, Edensor (2000, p. 326) argues, "an extension of tourism theatrical metaphors to include notions about the *direction* of performances, the *stage-management* of space and the *choreographing* of movement also helps to reveal the spatial and social controls that are mobilized to assist and regulate performance." He goes on to argue, whether tourism takes place in parks, museums, or cities, "these settings are distinguished by boundedness, whether physical or symbolic, and are often organized to provide and sustain commonsense understandings about what activities should take place" within them (Edensor, 2001, p. 63; see also Chaney, 1993; and Baerenholdt, et al., 2004). In the staging of tourism, symbolic boundaries are, arguably, more pervasive, as tourists engage varying degrees of overt and subtle landscape cues to interpret areas in which they might or might not be safe and/or welcome (Figure 5.2). In the analysis of pre-Katrina New Orleans, Atkinson (2004) illustrates the ways in which music was used to shape tourism spaces. She notes that where the music stops, so do the tourists. As a result, following the new waterfront developments, the city employed musicians to play along the area in the hopes that tourists would follow. Indeed, they did.

Subsequently, Edensor (2000; 2001) extends the stage metaphor to include notions of scenography and stage design, directors and stage managers, actors and intermediaries, and tourists as performers and audiences in the staging of tourism. Thus, he contends, tourists are not simply an audience to a staged tourism performance, but are an essential part of the performance, as "tourism constitutes a collection of commonly understood and embodied practices and meanings which are reproduced by tourists through their performances—in alliance with tourist managers and workers" (Edensor, 2001, p. 71). We extend this line of thinking in the next chapter, where we apply the rhetorical notion of the enthymeme to touring practice.

Figure 5.2 Staging National Identity: Flying the Dannebrog in Denmark

While the staging of tourism practices is more apparent in "enclavic" tourism spaces such as theme parks, elements which choreograph tourist movement and participation can also be found in heterogeneous tourism spaces, such as cities (Edensor, 2000; 2001). Thus, stage design extends to historicizing elements, such as "old fashioned" lampposts in urban tourism districts (Ashworth and Tunbridge, 2000) to signposts and markers that label points of interest and to guidebooks which direct tourists to the optimal viewing times of events. "Tourists are rarely left to draw their own conclusions about objects or places," argues Neumann (1988, p. 24), as they confront "signs, maps, guides and guide books [...] that repeatedly mark the boundaries of significance and value at tourist sites." Such staging and choreography are utilized in the performance of tourist identities. For example, backpackers tend to follow the same guidebooks, in particular *Lonely Planet*, but also signify their touristic identity through the use of costuming and props—a backpack, a journal, a camera, and postcard writing (Edensor, 2001; Jenkins, 2003; and Shaffer, 2004). These deliberate choices simultaneously enable backpackers to perform their roles as tourists and be read as a specific kind of tourist. Thus elaborations on MacCannell's stage metaphor have illuminated the mechanisms that encourage and regulate tourism performances; yet, the politics and power relations of these performances have received less attention in such studies. Returning to Goffman's concepts of framing and footing, as well as work in rhetorical theory, assists in bridging these conversations.

Tourism as Persuasion

It is necessary to briefly expand upon the concept of rhetoric as constitutive of both everyday and extraordinary life experiences, especially insofar as rhetorical situations saturate tourism experiences. Rhetoric is taken here to mean any form of symbolic interaction that exists to wield influence; persuasion can come in the form of text, speech, physical movement, or any other manifestation of opinion (Farrell, 1993; and De Certeau, 1984). Furthermore, manifestations of opinion need not be carefully formulated and expressed as such; texts not conceptualized with the intent to persuade are still argumentative (see Bitzer, 1968; and Austin, 1975). The case studies in this volume examine a variety of types of texts, as well as expressly stated and implied rhetorical messages. Theories of rhetoric, however, do not imply automatic audience acceptance of persuasive messages. Literature about the practice and faculty of judgment is especially important to the study of rhetoric, as it more fully accounts for the exchange between performer (a tour guide, for example) and audience (in this case, the tourists).

As Black (1965) argues, judgment is the object and end of rhetoric. Rhetoric's goal is, after all, to persuade, and persuasion would be impossible without the faculty of judgment. Gadamer (1983) echoes Black by contending that human existence is fundamentally interpretive. The study of judgment also provides occasion for noting how messy arguments tend to become, often intentionally

Figure 5.3 Tour Guide with Tourists at Spring Mill Pioneer Village, Indiana

so. Beiner (1983) examines such issues in his work *Political Judgment*, in which he argues judgment has problematically moved from the political to the private sphere, making people more prone to apathy. The end goal in matters of judgment (and hence rhetoric) is often some form of *phronesis*, or practical wisdom (Scott, 1998; Farrell, 1993; and McGee, 1980)—tourism produces an obvious and tangible version of this wisdom, inviting a transdisciplinary theorization of tourism and rhetorical studies (for example, see Garrod and Fyall, 1998; Ribeiro and Marques, 2002; and Pezzullo, 2003; 2007).

Tourism, Persuasion, and Activism

Working to uncover the mechanisms of tourism staging and performance, Pezzullo's (2003) work on toxic tours of Cancer Alley, Louisiana engages MacCannell's five stage process of sight sacralization. Because this employment of MacCannell's work is significantly innovative and has been tremendously influential, we investigate it in some detail here.

While Pezzullo agrees that such tours are choreographed and a series of stage techniques are employed, she argues that these performances, in particular, are more constitutive than "fake." Tours of Cancer Alley actually produce communities, particularly in relation to environmental justice activism, by contesting more conventional tourism representations. She begins by identifying the stage as the tour bus along with its various stops, its actors and directors namely the tour guides, local community representatives, and Sierra Club organizers, and the audience of tourists who often engage in environmental activism. MacCannell's first stage of sight sacralization is naming. In this case, the naming of an industrial corridor in Louisiana as "Cancer Alley" suggests not the area's physical or economic characteristics, but its impacts on those who reside there. Framing and elevation, the second stage, places a boundary around an area, putting it on display. For this stage, the author describes a persistent theme of the tour—to question what is worth preserving and by whom. Several chemical companies have purchased a number of properties along the corridor including one of the largest plantations in the state, a contaminated neighborhood turned "green space," and a church and cemetery. Stage three, enshrinement, is turned on its head with this tour, as the Ashland plantation estate, bought and renovated by Shell Chemical Company, is used to call into question authenticity and social justice. Photography and videography illustrate mechanisms of stage four, mechanical reproduction of the significance of the toured places. Social reproduction, which occurs when groups or places name themselves after their famous attractions, is the final stage of sight sacralization. Pezzullo does not identify this aspect of tourism staging in this case. In fact, conversely, the people who reside in this area are disappearing as a result of the area's high cancer rates and the rates at which industrial companies are purchasing properties. Thus at the center of this tourism performance is the contestation over corporate staging efforts to reduce the visibility of environmental and social impacts of an area's chemical industries.

The bodies constitutive of tourism create moments of persuasion and activism in turn. Toxic tour participants contest not only the visibility of environmental and human degradation—they contest those practices of degradation themselves. As Pezzullo (2007) notes, the very act of "being present" (p. 9) on a tour is an especially effective mode of advocacy. Though using very different language, Ebron (1999), Brin (2006), and Brin and Noy (2010) echo Pezzullo's argument. In the discussion of political tours in Senegal (Ebron, 1999), the Gambia (Brin, 2006), and Jerusalem (Brin and Noy, 2010), these authors explore how tourists are ideologically hailed and interpellated into particular ideologies (these ideological processes, as related to tourism and drawn from Althusser (1984), are discussed in Chapter 4). Ebron investigates a McDonald's-sponsored African American heritage tour, negotiating the corporatization and sanitization of oppressive histories and politics. In a slightly different vein and following the groundwork laid by Brin (2006), Brin and Noy (2010) explore the performative dimensions of place that occurred on a Jerusalem tour. The authors argue that through embodied performance, tour guides reaffirm ideological and political narratives in addition to publicizing historical events. Though this move might seem somewhat obvious, it is crucial to bear in mind the embodied and performative dimensions of the politics of place. Regarding Brin and Noy's study, an Israeli tour guide perhaps predictably relayed a narrative more informed by Israeli politics than Palestinian politics. While Israelis and Palestinians certainly construct complex, intertwined narratives about their highly contested lands, international tourists nuance their knowledge of the area through the act of touring. Tourism thus provides moments of continual reconstitution of place—especially for those less familiar with specific locations. This reconstitution is necessarily performative and persuasive.

Enthymematic Performances

> The problem with enthymemes…is that, if given carte blanche to fill in any proposition needed to make the inference structurally correct, we may insert assumptions into the text of discourse that the speaker or audience didn't realize were there, doesn't accept, or didn't even mean to be part of the argument...
>
> (Walton, 2001, p. 94)

A particular concept in rhetorical theory, the enthymeme, strongly contributes to a theorization of tourism. The rhetorical enthymeme has a sordid history of being poorly defined and generally confusing (Walton, 2001; Walker, 1994; and Bitzer, 1959). Here, we offer several different definitions of the enthymeme that will be useful in subsequent analyses. Bitzer (1959, p. 405) describes enthymemes as "premises [that] are asked for in order to achieve persuasion." He traces the importance of the enthymeme back to Aristotle's time, attempting to tease out its relationship to the syllogism while declaring it as the substance of persuasion. Perhaps more concretely, he declares that enthymemes substitute for question-and-answer sessions in oratorical performances—enthymemes are in effect "propositions which members of [the speaker's] audience would supply if [the speaker] were to proceed with question and answer" (Bitzer, 1959, p. 408). Walton (2001) and Walker (1994) offer two supplementary definitions. Claiming that the enthymeme is "an argument that has one or more premises, or possibly a conclusion, not explicitly stated in the text, but that needs to have these propositions explicitly stated to extract the complete argument from the text," Walton (2001, p. 93) explains that enthymemes are often called arguments with missing premises (though he prefers the term "nonexplicit assumptions" (p. 93)). Similarly, Walker (1994) writes that individuals naturally use enthymemes whenever they try to persuade each other. Conley (1984, p. 169) importantly notes several characteristics of the rhetorical tool—the two most central to this book are: (1) "the premises of an enthymeme are probabilities, not certainties," and (2) "the premises of an enthymeme are not simply statements of probable fact but reflect values and attitudes as well." While the first clarification relates more to performers, the second clearly implicates audiences—even if it does not focus on them.

Enthymematic interactions between guides and tourists are of special interest to this book. Bitzer (1959, p. 407) states, "the successful building of arguments depends on cooperative interaction between the practitioner and [her or] his hearers"; Walton (2001) discusses such rhetorical practices in terms of their implications for common knowledge, presumably among audiences. Enthymemes by their very nature assume particular kinds of shared knowledge. Such assumptions can (1) be made erroneously, and (2) be constructed through the use of rhetorical practices. Although the study of enthymemes has often been allocated to logicians, Levi (1995) argues the rhetorical context of any argument is every bit as important, at least socially, as its level of logical coherence. From a rhetorical perspective, enthymemes are about foregrounding an argumentative stance and encouraging personal identification with that stance (Walker, 1994). Along those lines, Finnegan (2005) suggests some of the importance of the enthymeme rests with the fact that it can fictitiously make context-driven, plastic premises appear as though they are empirical. Those premises become problematic (as quoted at the beginning of this section) for several reasons, not the least of which being that they encourage constant inference in ways that could quickly become detrimental to both independent thought and well-informed opinion. The relationship between tour organizers and tourists-as-captive-audience members creates a need for consistent rhetorical participation. The tour itself is necessarily enthymematic, and manipulatively so. While tourists might not always fulfill their enthymematic roles to the satisfaction of organizers and guides, their very presence already completes a series of important communicative events.

In the marketing of the Birthright Israel tour, the subject of Chapter 6, the enthymematic and performative notion of "the gift" is one of the most pervasive tropes and thus it deserves at least a cursory theorization. Though he writes about "primitive societies" in terms no longer used today, Levi-Strauss (1996) fundamentally argues that there are social systems in which economies are based not on profit but rather on gifts. Furthermore, the principle of reciprocity is inherent in the gift; gifts are not simply given as gifts, but are laden with meaning and the presumption that the gesture will be returned in one capacity or another. Mary Douglas, in her introduction to Mauss' *The Gift*, argues there are no such things as "free" and/or "pure" gifts (Adloff, 2006). In addition to being an exchange of goods, Mauss (1925) also argues that gift exchange can be a transfer of identities. Sahlins (1996, p. 32) notes that in generalized reciprocity, exchange can be based upon "kinship dues," meaning that the ability to take part in the exchange could be reliant upon a type of kinship. Tourism thus fosters the exchange of goods, people, and to a certain extent, cultural practices. More deeply, cultures are constructed as being able to be transferred and transplanted, if only metaphorically, through tourism.

Tourism as Performance of Place

Despite efforts to choreograph and direct tourists, tourism's dramaturgical landscapes cannot *determine* performance (Edensor, 2000; 2001). Indeed, most tourism spaces are more heterogeneous than homogenous, suggesting tourists are just as likely to encounter non-touristic social performances. Because these spaces combine tourist facilities with local businesses, public and private institutions, and domestic and guest housing, tourists mingle with locals. As such, heterogeneous spaces provide a stage upon which tourist performances of place "may be performed alongside the everyday enactions of residents, passersby and workers" (Edensor, 2001, p. 64) (see Figure 5.4). As a result, the stage metaphor does not account for the spontaneous moments of being a tourist. By turning towards performance theories, there is room for new metaphors based more on agency—the being, doing, touching, and seeing of tourism (Cloke and Perkins, 1998; Coleman and Crang, 2002; Baerenholdt et al., 2004; and Larsen, 2008). Place is not a static container, but it is made through embodied experiences (Coleman and Crang, 2002; Baerendoldt, et al., 2004; and Rickly-Boyd, 2013b). Therefore, the performances of tourism places are accomplished by sets of discourses and texts, bodies and objects, affects and percepts, and technologies and mediums (see Coleman and Crang, 2002; and Baerenholdt, et al., 2004). Notions of place are validated not by simply seeing its symbolic features, but through tourism performances that include spontaneous actions, senses, and emotions, along with signs of the everyday (Rickly-Boyd and Metro-Roland, 2010). It is through tourism moments and embodied practice that tourists are able to have a sense of experiencing "real" places.

Figure 5.4 Urban Tourists and/or Locals in Montevideo, Uruguay (by Nicholas Wise)

Buchmann, Moore, and Fisher's (2010) study of *Lord of the Rings* tourists in New Zealand illustrates the interweaving of these themes. Despite the fact that tourists are motivated by the film series developed from Tolkien's works and that the tours have aspects of mediated hyperreal simulacra, the authors found that tourists repeatedly expressed a sense of an authentic experience. In particular, they note being able to connect to the themes of Tolkien's stories—fellowship, adventure, and sacrifice—through various aspects of the tour. A sense of fellowship comes from forming a touring community with others of common motivation, and this extends to the guides who lead the nationwide tours—"someone who is personally touched by the subject and willing to not only 'join the fellowship' but also to lead it" (2010, p. 242). Furthermore, they experience the adventure of travelling far distances to places only imagined. In fact, being in the landscape that was used as background to portray the Tolkien stories on film, tourists describe engaging their minds and their bodies together. They are able to experience a material place, its terrain and climate, but also use their imaginations. This embodied experience is essential to the performance of place. Thus the authors argue that embodiment "helps counteract feelings of surreality that many tourists reported when arriving" (p. 241). Such studies of the embodied practices of tourism remind us that tourism is not just about sightseeing, but more importantly about journeys that allow one to *be* in place. As such, tourism performances of place are not bounded to specific locales, but rather extend constellations of places.

Non-Representational Theory

Human geography has recently taken a "performative turn," with greater focus on the enactment of space, the body, and embodied experiences (Thrift, 1996; 1999; 2000; Winchester, 2005; Wylie, 2007; and Anderson and Harrison, 2010). While performative methodologies, those that attend not only to the agency of humans but also things and accept "unspeakable" elements as crucial (Clark, 2003; Harrison, 2006; and Laurier and Philo, 2006) are still inspiring debate, analytical perspectives informed by performance theory are becoming more common. Much of this analytical perspective in human geography has come under the term "non-representational theory (NRT)," coined by Nigel Thrift (1996; 1999; 2000), who has brought together a number of philosophical perspectives in order to encourage social scientists, and geographers in particular, to re-evaluate their emphasis on representation and to examine, instead, how space and time emerge through embodied practice. NRT, in Thrift's own words (2008, p. 2), is a "geography of what happens." Dewsbury et al. (2002) point out, however, that this is not an anti-representational approach, but is intended to push beyond examinations of representations, to be "more-than-representational" (Lorimer, 2005). Specifically, Laurier and Philo (2006, p. 353) note this perspective is interested in "matters that mark the end of representation: things, events, encounters, emotions and more that are unspeakable, unwriteable, and of course, unrepresentable."

Non-representational theory, according to Dewsbury, et al. (2002, p. 438) is "characterized by a firm belief in the actuality of representation [...] not as a code to be broken or as an illusion to be dispelled rather representations are apprehended as performative in themselves; as doings." The act of representing, whether by speaking, writing, or creating, argues Wylie (2007, p.164), is understood to be "in and of the world of embodied practice and performance, rather than taking place outside of that world, or being anterior to, and determinative of, that world." Thus tourism offers countless examples of the ways in which representations are performed and inform performances. Photography and photographs offer just one, albeit pervasive, example of the ways in which representations are performative.

In the latest edition of *The Tourist Gaze* (2011), Urry and Larsen further incorporate the performative aspects of photography into the concept of the gaze. Examining the performativity of photography, the authors attend to the agency of the materialities that (re)produce photographs and embodied practices that accompany photography; in addition, they illustrate how photographs organize particular gazes. In tracing the history of visuality and tourism, Urry and Larsen describe the various technologies that have shaped tourist gazes over time, from the use of Claude glasses (tinted mirrors) to the invention of the camera. The increasing availability of the camera and the use of photographic images in tourism advertising have worked hand-in-hand to inform the role of photography in tourism. The Kodak company, in particular, "re-made and re-scripted photography as a leisurely family-centered performance" (2011, p. 170), and in further informing the tourist gaze, the authors note that the desire, then, to capture "Kodak moments" becomes, in effect, "a search for the photogenic" (p. 178). Yet, they contend, "photographic performances are always *more*-than-just representational" (p. 209). Photography is an embodied practice performed by those on both sides of the lens. The photographer does not simply point and click, but bends, squats, and leans to produce the preferred angle, and similarly, those photographed pose and posture.

Further, non-representational approaches understand landscape not as an "inert background or setting for human action, nor is it understood as simply a pictorial or discursive form" (Macpherson, 2010, p. 6), but as "perception in motion" (Wylie, 2007), "process" (Rose, 2002), "practice" (Cresswell, 2003; and Wylie, 2007), or "dwelling" (Ingold, 1993; 2000; and McHugh, 2009). This is not a dismissal of the representational qualities of landscape, but a reinfusion of agency. Thus, Rose (2002, p. 457), states NRT foregrounds the notion of "how the landscape 'comes to matter'—how it comes to be relevant through practice." This theorization, therefore, has considerable implications for greater understanding of the tourism encounters.

Wylie's (2005) account of his walking tour of the South West Coast Path of North Devon, England illustrates the power of non-representational approaches to addressing how landscape is performed; how it "comes to matter." Using first person narration, Wylie emphasizes the body as he explores the performative orderings of self and landscape of just one day's walk of a three-week, 200-mile journey. The 4th of

July 2002 begins "in the woods" outside Clovelly, with experiences of nervousness and anxiety upon entering the immensity of the tangles of the woods' branches, leaves, and undergrowth. Stopping in the woods, however, reveals the intimacy of "this tree, these branches," and with it a transition from a sense of solitude of walking in a wooded landscape to an enmeshment of self and landscape. Although the woods end as he moves into a valley, the "thickness" does not. Wylie now follows just a small, mud path through wet, thick blankets of ferns—"[l]imbs and lungs working hard in a haptic, step-by-step engagement with nature-matter. Landscape becoming foothold" (2005, p. 239). Despite what may seem like a description of "being-in-the-world," Wylie argues of this experience, "[w]alking does not embed the self 'in' landscape, nor does it put in motion a relation in which some auratic sense of self and place emerges" (p. 240). Thus, he suggests in this moment he is as much "in" the landscape as up against it. Escaping the fern-choked gullies, he approaches the coastline and notes a change in his perception of the ocean. "After a few days walking, the sea stops being, as it were, the edge of the land, the end of one thing and the beginning of another. It becomes instead a sort of encircling element. As a coast walker I began to be very aware of being on an island [...] the sea seemed indifferent to the land" (p. 241). From here, the path curves and moves through the Smoothlands, a series of valleys along the coast. While he argues his cultural baggage of Western aesthetics suggests he should appreciate this "resplendent landscape" of hills and valleys, because of the vast distance he now looks upon he instead experiences a sense of sublimity—awe and dread—and he is left with self-doubt and body doubt about undertaking the project.

> [A]s a walker becomes chaffed, jarred and footsore, so the landscape no longer takes shape as a set of readily affording surfaces for purposive and smooth motion. Instead, the world contracts and the subject splits. [...] The bone pain of walking is realized in an aching halo of landscape, with the ground immediately beneath your feet and the slope climbing above and the coast unspooling relentlessly ahead. Pain occurs neither 'in me' nor 'in that'—the externalized body—but 'between me and it', in this step, this next step. And so the landscape emerges as malignant. (p. 244)

Wylie's travelogue-style account offers an exploration of the changing manifestations of self and landscape through his narration of a day's walking by utilizing a post-phenomenological understanding of self and landscape in practice. A post-phenomenological perspective for tourism considers not just the moment of the tourist's experience, but the context through which it is experienced culturally, socially, and bodily.

Embodied Practice

Attempting to overcome the mind body dichotomy of Cartesian thought, Merleau-Ponty argued that experiences exist between the mind and body

(Leavy, 2008). However, non-representational theory is actually considered post-phenomenological because it extends beyond the body to how and why our bodies are put into motion or emotion. Thus, the context of embodiment is an important aspect of this perspective. Leavy argues (2008, p. 346), "the body is not viewed as an object, but rather as the 'condition and context' through which social actors have relations to objects and through which they give and receive information" (see also Grosz, 1994). Bodies are performances that occur in conjunction with landscapes, objects, and contexts, informed by habits, sensations, and materiality (Macpherson, 2010) (see Figure 5.5). The body is "always both subject and object...observer and observed, seer and seen" (Wylie, 2007, p. 151). Indeed, Crouch relates bodies and space in tourism, suggesting, "[l]ike bodies, space becomes inscribed with significance, and both are powerfully deployed in tourism" (2005, p. 26).

Figure 5.5 Embodiment of Place: Surfing Bodies in Hawaii

Performance is an enacting of space. So while tourism studies originally used theatrical metaphors to understand the structural and ideological dimensions of tourism, stepping beyond the boundaries of the stage to investigate the performance of tourism, by hosts and guests, offers a way to access active, "in the moment" experiences of tourism places. As Crouch (2002) argues, "[t]he subject doing tourism makes lay knowledge through a complexity of awareness that is immediate, diffuse, and interactive [...] we 'know' places bodily and through an active intersubjectivity" (p. 214).

Senda-Cook's (2012) study of hikers in Zion National Park provides an example of more recent work in rhetoric and tourism studies to integrate embodied experience, rhetoric, and performance theories towards a more robust understanding of the relations between tourist experience and tourism site. Her research on outdoor recreationalists illustrates that assumptions about authenticity are constructed through embodied, repetitive performances and, as such, can be produced and sustained or undermined through rhetorical practices at the park. In particular, she considers tourist articulations of "quality of experience" in relation to lived experience, and more specifically, experiential degradation that results when expectations of authenticity and its performance are not met by other tourists' practices. These include contestations among tourists with regard to on-trail hiking as responsible environmental stewardship and off-trail hiking as more spontaneous and thereby authentic, as well as rhetorical practices and materialities which communicate self-sufficiency, preparedness, and appreciation of risk. Thus, Senda-Cook weaves together several themes discussed in this chapter, along with previous theoretical chapters, and more importantly, demonstrates the networks of politics and performances that overlap when tourists are in place, among each other, and always already interpellated into multiple hierarchies of ideologies.

Conclusion

Tourism is not "merely *like* a performance," Bowman and Pezzullo (2009, p. 193) contend, "it *is* a performance insofar as the site is often composed of live bodies engaged in acts that are put on display for tourists" (emphasis in original; see also Desmond, 1999; and Kirshenblatt-Gimblett, 1998). Likewise, tourists are performing identities and enacting place. Thus, while dramaturgical metaphors have been useful for examining the structural/functional and ideological aspects of touring, performance-centered tourism research examines actual, bodily enactment. Yet, in so doing, such research attends to the performance of tourism places and also considers the continual semiosis of tourism in which interpretation and action are dynamical. As such, it accounts for ideology in tourism sites as not merely determining space and action, but continuously requiring social practice, that is, to be put into practice by tourists, locals, and tourism practitioners. Performance-based theories, thus, offer a way to examine the tourism site and the tourist experience within a single framework.

PART II
Tourism Performances

Introduction

Whereas Part One is intended to demonstrate the depth and breadth of each of the primary theoretical constructs—ritual, semiotics, ideology, and performance—for the study of tourism, Part Two aims to illustrate these theoretical relations through four case studies of tourism performances. Each case study foregrounds performative aspects of tourism while also attending to one or more of the other theoretical constructs of ritual, semiotics, and ideology. In maintaining a geographic perspective on tourism, place is a central concept. Place is not simply a destination or a location that can be bounded, place is performed by tourists and locals, tourism practitioners, and those who labor in the industry. In other words, place is enacted through touristic practices. Considering the manifestations of the performative turn in tourism studies thus far, we can identify the following trends that have been brought the fore:

1. Embodiment: The performative turn adjusts the overt focus on the visuality of tourism to attend to the role of the body, sensuous encounters, emotionality, and affect in tourism. It suggests that tourism is about more than seeing or gazing, but also concerns the embodied practices that perform places, in which sightseeing may be but one performance.
2. Agency: While many aspects of tourism performances are *pre*formed, attention must also be paid to the agency of all actors. With the performative turn, dramaturgical metaphors for understanding the ways places are scripted, staged, and choreographed are expanded to include the power relations of the agents involved. Further, a number of non-human agents are considered essential to tourism—cameras, tour buses, passports, etc. These materialities have agency and thereby affect tourism performances and mobilities.
3. Representations as doings: The performative turn challenges representational and textual investigations of tourism, advocating for attention to actions that (re)produce representations as well as disregard, challenge, or ignore them. It highlights the ways representations are made meaningful through the actions they inform and how representations are changed accordingly.
4. Mobility and Connectivity: Tourism does not just happen in places, but places are performed through tourism practices. As such, tourism is about

> mobility: the mobility of people, institutions, objects, images, and things. Mobility forges connections between places, as tourism destinations are points in networks of movement.

Each of these areas of interest has considerable implications for understanding the role of place in tourism. A performative approach to place accounts for the role of the body in the production of space and therefore the significance of embodied practices to the enactment of place. Embodied enactments of place highlight the varying and often unequal agencies acting in place, attending to differing and often contested ideologies that continually (re)produce representations. These enactments of place, then, are not bound to particular, insular spaces but cut across and connect places. Thrift (1999, p. 310) states, "places are 'passings' that 'haunt' 'us,'" that is, places take shape only in their passing—they haunt us and we haunt them. Thus, places are articulations of presence and absence, and as such, can never be pre-ordained (Thrift, 1999). Indeed, in the context of tourism, the places performed through touristic practices stay with us and we remain connected to them, as they become points in constellations of places.

Considering place as performative also opens up a multiplicity of potentialities for what places can do. It is through our performances of places that places take on meanings and become affective. In this affective spirit, touristic practices have the potential to be transgressive by challenging the ideologies of places. Of course, in many cases, tourism performances reify existing, naturalized ideologies. In the next chapter, we take up such issues. The case of Birthright Israel, a tour for young Jews to the State of Israel, is exemplary of ways in which ideology and persuasion are performative, as well as *pre*formed. While this tour of Israel as (Jewish heritage) place is presented as a "gift" to participants, engaging a rhetorical critique illustrates that these circumstances also create obligations of its participants. Each stop on the itinerary thereby corresponds to ideological agendas that participants have restricted agency to challenge in their individual performances. As a result, this strictly regimented tour fuses land and emotionality by repeatedly juxtaposing decisions of "life" and "death" at specific points in the landscape as a means to foster a stronger sense of identity between its participants and the nation-state. To understand ideology as performative, agency comes to the fore.

Identity politics are inherent to the production and performance of place. Tourism, in particular, offers a forum for the projections of carefully crafted place identities (Light, 2000; 2001). Indeed, in choosing which sites to authorize as emblematic of place, tourism organizations construct place myths in which individual sites act referentially. As a result, the deconstruction of tourism marketing has been a central practice of tourism scholars, as it reveals the power relations and ideologies that constitute distinct processes of othering. The performative turn in tourism studies, however, extends such analyses to ask questions of the practices that maintain or challenge particular representations of places and the ways in which representations influence touristic practices. In Chapter 7 we work across these approaches by examining the case of tourism promotion in Denmark. Our investigation of a promotional video illustrates the interlacing of semiotics and ideology in constructing place myths. The identity politics of crafting a marketing device that conveys "Danishness" is thus also necessarily a process of othering. By determining what is "Danish," the tourism organization VisitDenmark is also, implicitly, stating what is not Danish. Yet, such marketing campaigns work on the presumption that viewers will be able to distinguish "Danish" from "Scandinavian," "European," or even just "foreign" and will be enticed to visit Denmark. We find a number of instances in which this fails to be the case, so that potential visitors are not convinced, and are in fact confused by the choice of imagery. In those cases, a desire to visit is not induced. In other words, the representations fail to perform. In the second part of the investigation, we look specifically at on-site materials from a series of sites across Denmark to assess the ways in which these materials influence and even limit interpretation. Comparing translations of Danish materials into other languages, notably English and German, in most cases the information offered is quite similar. However, in several instances, severe degrees of slashing, slippage, and misperception are identified. In other words, information has been omitted, miscommunicated, or misunderstandings of cultural perceptions have occurred in the translated materials. This has considerable implications for tourists' performances of these sites. As representations of particular tourism sites, they influence touristic practices and interpretations. Such interpretative materials work to frame potential performances by offering only select information.

Promotional devices can also be understood as representations and doings. Such devices necessitate an investigation of the ways guidebooks, websites, and brochures are integrated into individual touristic performances of places. Are the "must see" sites indeed "seen" and as a result their representational power reified, or are they overlooked, that is, not performed? Incorporating such questions into Graburn's theory of tourism as a ritual, in Chapter 8, we illustrate the generalizability of the ritual practice of tourism while also attending to individual enactments of place and the experiences of communitas that develop in the performance of tourism representations. While Florence, Italy is the primary focus of this chapter, we make comparisons to autoethnographic touristic experiences in the, arguably, similarly touristed city of Venice. Drawing on the work of Tuan (1974), relations are observed between sightseeing performances, communitas, and notions of place. In the case of Florence, MacCannell's conceptualization of sightseeing as modern ritual is situated within Graburn's more encompassing theory of tourism as secular ritual. In so doing, the significance and circumstances of communitas experiences while touring are highlighted. Ritual performances of Florence do enact it as place; the city becomes a passing that haunts. Yet, these experiences of community and place are fleeting, suggesting Florence remains more of a "public symbol" in the Tuanian sense of place. Examining the autoethnographic experiences of Venice, however, illustrate Tuan's "field of care"; not only has sightseeing been performed there, but communitas experiences have been transformed into extended community networks. Indeed, comparing these two cases of tourism within ritual frameworks foregrounds the performativity of place.

The performative turn in geography, specifically, argues place is made through connections, not eroded by them. Since this framework is well-suited for considering touristic performances of place, it then suggests questions of what implications the performative turn might have for some of the other foundational concepts of tourism, namely "home." Must tourism necessarily take place away from home, for specific durations? Lifestyle travel and lifestyle mobilities problematize such assumptions. In Chapter 9, the notion of the (hyper)mobile home is a central concern. As a result, the investigation of a particular form of lifestyle travel/mobility integrates tourism studies and mobilities studies. Full-time, non-sponsored, traveling rock climbers are the focus of that chapter. Referred to as lifestyle rock climbers, these individuals have given up sedentary residences to travel continuously in the pursuit of the sport. In so doing, most transform a vehicle into a mobile abode. Interrogating the ways in which home is performed on the road, then, necessitates attention to performances. Home is not about the structure in which one resides, but it is produced through embodied practices of everyday rituals and habits, in the social relations that connect individuals to communities, and through feelings of security and familiarity. In the case of lifestyle travel/mobility, home is not left in order to tour but home is continually practiced while traveling. And, indeed, so is place.

Place has a different role in each of these cases. Drawing from various situations, attention to the ways in which place is performed ties together the theoretical constructs of ritual, semiotics, and ideology while calling attention to the roles of embodiment, agency, representations as doings, and mobility and connectivity. In tourism performances of place, each is active, yet comes to the fore at different times and in different circumstances.

Chapter 6
Performing Tourist Ideologies: Israeli Birthright Tourism

Introduction[1]

Tourism, in its enactment of rituals through performance, extensively utilizes persuasion to inform opinion and adjust thought. This chapter analyzes the expressed and implied ways persuasion can systematically build ideology in the tourist experience. More specifically, it seeks to examine how the rhetorical qualities of the American program Birthright Israel work to influence participants' ideas and feelings about the State of Israel and their personal Jewish identities. Birthright Israel (also referred to as simply "Birthright"), a free, ten day trip to the contentious state, is offered to young adult Jews around the world as a "gift." The tour has a fairly explicit agenda—to (re)connect young Jewish adults to both their heritage and a nation that generates significant controversy—but it engineers a variety of implicit effects upon participants. I seek to identify particular, inherently persuasive methods used on Birthright Israel tours that extend beyond traditional speech in order to influence tourists. In doing so, it is argued that rhetorical studies are indispensable to the study of tourism.

Justification for this chapter lies in the belief that rhetorics of tourism are understudied, especially as they relate to individuals' abilities to render place-based judgments. Rather than accepting the idea that tourists are influenced by every persuasive message that comes their way, I wish to complicate the notion of a rhetoric of tourism. The tour creates an extraordinarily effective (and affective) site for concentrated, persuasive experiences, and individual trips can be used as case studies to examine how performance and persuasion intersect. Particularly, I focus on the efficacy of the enthymeme in creating a tourist narrative that allows places to be mutually constituted and gifted (for a definition of the enthymeme, see Chapter 5). With the case of Birthright Israel in particular, there are many implications of this theoretical maneuver—namely, Birthright exemplifies how touring can be an instance of bringing a place to a person, even when it appears as though the person is being brought to the place.

---

1 This chapter is based upon research and participant observation conducted by Lisa Braverman. Because of the highly personal and individualized nature of this work, the authors have chosen to retain the personal "I" as opposed to embracing a collective "we."

Theoretical and Methodological Underpinnings

This chapter is an application of the concepts of the enthymeme, performative persuasion, and the gift outlined in Chapter 5. While this research is firmly rooted in the experience of the tour, it also nuances dialectical tensions between ideology and agency. As argued in Chapter 4, ideology plays a strong role in virtually every aspect of tourism. Sites are often (though not always) selected and marketed by host cultures, and the entailments of such choices ideologically guide tourists. Top-down notions of ideology are not sufficient to explain diverse experiences of touring; a strong performative component animates tours. As this chapter argues, tourism performances are also far from top-down. Embracing a theoretical construct in which guides resemble speakers and tourists audience members, tourist bodies are themselves used to complete ideological arguments.

This chapter employs the methods of participant observation, interviewing, and discourse analysis. Acting as a participant observer gave me the opportunity to both rhetorically analyze the tour first hand and contextualize many interviewees' statements. As Dewalt and Dewalt (2002) note, "a good deal of what we learn in the field is tacit" (p. 68)—this aspect of fieldwork has the potential to be both beneficial and detrimental to projects. Observation techniques, and hence an ability to cultivate unspoken knowledge, is necessary to this research because it grants the critical opportunity to extend analysis beyond that which is immediately apparent or readily legible. Although participant observation can be criticized for its inherent subjectivity, it is useful in the sense that it allows a researcher to cognitively and physically transcend ideological and spatial barriers, quite literally "going elsewhere" (Bernard, 2006). My most nagging ethical concerns arose in this portion of the research since I engaged in a low level of deception.[2] Bernard (2006) notes, however, that different research situations call for varying levels of deception and concealment, and that deception is often necessary in studies that involve in-groups and out-groups. As a tour participant, I took copious field notes but refrained from formal and informal interviewing—while on the tour—so as to avoid impacting both participants' independent thoughts and their judgments of me. Upon returning from the field, I coded and analyzed my field notes along with the interview data I accumulated both before and after the excursion.

I interviewed two people both before and after they participated in the tour, four people after the tour only, and two people who were considering Birthright but had not yet participated. While interviews are regarded as a legitimate and

2 I say "low level of deception" because initially, I had not planned on disclosing my status as a researcher on the tour (and I never did, fully). Because of the intimate nature of the experience, however, it was impossible to hide the fact that I was taking field notes. Generally my fellow tourists were so disinterested in me, their lines of inquiry never extended very far. By the end of the trip, I had told them that I was a graduate student doing research on tourism and this trip would help me with my thesis–none of which bothered them in the slightest (at least visibly).

foundational way to gather information beyond a first person perspective (Kvale, 2008; Dewalt and Dewalt, 2002; and Bernard, 2006), they are also seen as sites of disproportionate power relationships and frequent misunderstandings and misinterpretations (Briggs, 1986). Interview subjects were selected using a snowball sampling technique. Snowball sampling, of course, has clear disadvantages due to the way in which it privileges members of a population who know the researcher—a small sample of a much larger population—yet there were strong supporting reasons for employing this technique. Previous attempts to stratify and randomize the sample revealed that I would likely have access to fewer trips and trip organizers with a strictly self-selecting pool than a snowball sample. Attempts to contact strangers through the Hillel[3] center on various campuses produced only participants who traveled on the Hillel tours; with a snowball sample, however, each of my interviewees participated on a different trip and four trip organizers were represented.[4]

The interviews I conducted for this project were semi-structured and in-depth. They were transcribed and treated as rhetorical texts, both informing and complicating my argument. Within the context of this research, interviews are not regarded as slivers of a composite "Truth," but rather as critical texts that are worthy of examination and meant to complicate the results of participant observation.

Keeping with the theoretical and analytic nature of this project, I performed discourse analysis on written and oral media related to Birthright Israel. Discourse analysis focuses upon language-in-use as opposed to a quantitative measuring of coded themes often referred to as "textual analysis" (Barker and Galasinski, 2001; and Brown and Yule, 1983). The two methodological forms appreciate the power of language, yet do not in and of themselves "prove" anything (Carlson, 1985). I included information specifically about media with which the average Birthright tourist would come into contact.

Birthright Israel

Birthright Israel is a ten day, all expenses paid trip to Israel that is marketed as a "gift." The only eligible recipients are Jews between the ages of 18 and 26.[5] Everything from the roundtrip plane ticket to each meal is paid for; the only obligatory fee is a fifty dollar tip for the guides. According to the program's website,

---

3 Hillel is the nationwide, college campus Jewish student life center.

4 A trip organizer is a sub-group that operates tours under the umbrella organization of Birthright Israel.

5 In this context, applicants to the program are interviewed before being allowed to take part in the tour. Everyone is questioned about basic Judaism and involvement in local religious communities, and their answers—along with written application questions that ask about family history—determine whether or not they qualify as eligible Jews.

hundreds of thousands of people have participated in the tour and the trip's stated aim is to "[…] provid[e] the gift of first time, peer group, educational trips to Israel for Jewish young adults […] in order to strengthen participants' personal Jewish identity and connection to the Jewish people."[6] Birthright materials stress the importance of fostering and maintaining a strong Jewish identity; throughout the tour itself, there are also a variety of pressures to date and marry someone Jewish (maybe even someone on the tour!). The religious and community-building aspects of the experience are infused with the political; the tour exists not simply to facilitate heightened identification with Judaism, but as the name "Birthright Israel" suggests, the program explicitly encourages multifaceted support of the state. Pro-Israel arguments on the tour take all kinds of traditional and nontraditional forms, including but not exhaustive of explicit guiding during site visits, community encouraged valuations of life and death, various degrees of Israeli military presence, and nonverbal communication.

In terms of their itineraries, the trips have fairly little variation. Taglit Birthright Israel, the "umbrella" organization, oversees approximately fifteen "trip organizers" in the United States and Canada.[7] Some trip organizers may place more emphasis upon outdoor experiences or community service, for example, but all groups make certain stops such as Masada and the Bedouin tents. A sample itinerary provided by Birthright Israel reads:

Day 1: Arrival at Ben Gurion Airport
Day 2: Tel Aviv-Jaffa: Tel Aviv, Old Jaffa, Rabin Square, Independence Hall
Day 3: Jerusalem: Western Wall, Old City, Mega Event
Day 4: North Coast: Meeting Israelis, Caesaria, Acco, Haifa
Day 5: Galilee: Zvat, The Jordan River
Day 6: Galilee: Golan Heights
Day 7: Shabbat (Sabbath)
Day 8: The South: Beit Guvrin (archaeological site), Bedouin Village, Camel Safari, Ramon Crater
Day 9: Dead Sea: Masada, Ein Gedi nature reserve, Dead Sea
Day 10: Central Israel: Modern Jerusalem, Mt. Herzl, Yad Vashem, Saying Goodbye

Itineraries are, of course, subject to change based upon security threats and weather conditions. In part because all trips are so similar, the marketing and recruitment for each experience is relatively homogenous.

---

6 This information was obtained from the program's website (http://www.birthrightisrael.com/site/PageServer?pagename=about_faq#23, accessed November 2008). For a full description of the motivations and background behind the Taglit Birthright Israel Program, see Shaul Kelner (2010), *Tours that Bind: Diaspora, Pilgrimage, and Israeli Birthright Tourism*, New York: NYU Press.

7 Information in this paragraph is drawn from the program's website.

Pre-Trip: Marketing, Recruitment, Signup and its After-Effects

Although at its inception Birthright marketed itself widely and explicitly—I heard about it in 2000 or 2001 on a radio ad—the need for it to do so in recent years has virtually evaporated. The people with whom I spoke uniformly came to Birthright by virtue of hearing others talk about it—almost all participants knew someone who had gone on the trip, and college students also heard about the tours on their campuses through Hillel organizations. This lack of overt marketing, due at least in part to the overwhelming success of the program, perhaps inadvertently heightened the social allure of Birthright. Before the trip, both program and nation were constructed in imaginary terms—in the truest Andersonian (2006) sense, prospective participants were called upon to envision both each other and the country they wished to visit.

On the program's website (when I was signing up), the slogan "Your adventure. Your Birthright. Our gift." was presented in a large font on the top banner of the page. Photographs in the banner keep rotating, but always include two people who appear to be having fun together. Three pairs of individuals were shown in succession: one pair of females, and two pairs that feature a male and female. At least one set of people appears to be dating. The tour is immediately coded as a social endeavor. By showing pairs of tourists instead of individual travelers, organizers subtly convey that one's participation in the tour means she or he will not be alone—the promise of collective fun, maybe even romance, awaits. Additionally, the aforementioned slogan intertwines the tourists, philanthropists, and organizers. The first two statements, "your adventure" and "your birthright" concentrates a focus upon young "gift" recipients, yet the third statement ("our gift") directly references the money those tourists are not paying—and hence the people who did pay in order to make the experience possible. While the tour is indeed a gift, it is not a gift without sacrifice. As a "present," the tour also forms a tenuous bond between strangers (since the tourists and benefactors, in most cases, do not meet). Participants are not asked to acknowledge their benefactors in any way, yet I have continued to receive almost weekly emails encouraging me to donate money in the interest of future tourists. Irrespective of its various iterations, the rhetoric of the pre-tour period plucks prospective travelers from solitude and removes the choice of whether or not to be social.

As demand for the tour has increased, registration—a pre-trip ritual all its own—for many trip organizers has become more and more chaotic. Knowing that I needed to secure a spot on a trip for research purposes, I joined thousands of other eligible Jews on Taglit's website on February 19, 2009, the first day individuals were allowed to sign up for summer tours. The server crashed in the morning, and it was late afternoon before I could get through to actually register.[8] Signing up

8 Registration does not imply acceptance to the program. Spots are filled on a first come, first serve basis, with priority given to individuals who had been waitlisted during the previous season. Once a spot has been allocated for an individual, a brief phone "interview"

was simple and did not take much time; I filled out basic information, scanned and emailed copies of my passport, and paid the mandatory yet refundable $250 fee. Interestingly, the security deposit/fee, promised to be refunded upon our successful completion of the trip, was not going to be credited to our bank accounts or credit cards. Instead, we were told we would receive a prepaid debit card. I thought this was somewhat odd, but disregarded it until I received the card in the mail several months later. Intricately designed, it featured the Birthright Israel logo in the upper left hand corner, and had a miniature collage that spanned the entirety of the card. Two more small groups were shown having fun, the bottom right corner had a street sign pointing in the direction of Jerusalem, and a rather large photograph of the Israeli flag nestled itself in the middle of it all. We were, therefore, not (permanently and monetarily) paying for our trip, but if we wanted to use our money, we were required to become card-carrying affiliates of the Birthright organization—which could easily be misconstrued as being supporters. I tried to use my debit card in select contexts, ones in which no one else would have to handle it. Once, I gave the card to a retailer who looked at it for a significant amount of time and shrugged before returning it to me. I was serving as a marketer for the trip but did not fully intend or want to do so.

This pre- and post-trip linkage, made manifest in the organization-issued debit card, is a clear example of the embodied, enthymematic gift. More broadly, it is symptomatic of a form of tourism that relies upon tourists for the completion of argumentative premises. In very structured ways (i.e. by paying the security deposit), I became tied to Taglit. In order to complete the process that was part of the original contract—that is, having my deposit refunded—I was required to use yet another "gift" that had been bestowed upon me (the card). On the whole, Birthright was very clear about its intentions to engage participants in a series of conversations that were meant to extend long into the future. The methods involved in initiating and maintaining those conversations, however, were at times disturbing, radically engaging, and almost always proceeded without consent.

The Tour: Rules, Rhetorical Methods, and Invitations to Inhabit

> …tourism is a safe place for practices that are contested in other spheres.
>
> (Bruner and Kirshenblatt-Gimblett, 1994, p. 448)

Within this section, I propose two interrelated types of rhetorical maneuvers that were used on my tour and my interviewees' tours to influence thought. These categories are meant to be neither simplistic nor mutually exclusive, and were

---

is required to confirm that you are in fact Jewish. During this interview, I was largely asked about the religious background of myself and my parents, if we were members of a synagogue (and which one(s)), etc.

derived from my participant observation experience. I regard my interviews as texts in and of themselves, and I quote the individuals I interviewed in order to further complicate, contextualize, and deconstruct Birthright Israel experiences.

The Persuasiveness of Movement

My group's tour began in Tel Aviv, and the first site we visited constructed an initial, serious, and deeply enthymematic frame of Israeli statehood that the organizers expanded upon as the tour progressed. Independence Hall, located in Tel Aviv, is the place where David Ben Gurion declared Israel's statehood in 1948. It has several floors, but we only saw the first. Throughout the course of the tour, we saw four rooms: the lobby, a small movie theater, the independence declaration room, and restrooms. After waiting briefly in the lobby we were escorted to the movie theater, where a guide showed us a film about the creation of Tel Aviv. The film argued that Jewish families were setting up suburbs of the Arab town of Jaffa; that the land, in effect, belonged to nobody. At the film's conclusion, the guide showed us a picture of the Zionist figure Theodor Herzl, and spoke about Herzl's portrait in the "other room." This segue provided us an opportunity to change locations and move to the second room—the place where Israeli statehood was officially declared.

Independence Hall is an effective miniature case study of the practice of progressive, enthymematic framing (for more on framing, see Chapter 5). Sitting in the movie theater, we were observers to the early days before Israeli statehood; once we moved into the room of the declaration, however, we were fictive participants in the creation of a nation. The "room of declaration" was set up to look as it had then—clearly it had been updated with a contemporary sound system, but the long table at the front of the room and the 200-300 chairs that populated the surprisingly little amount of floor space were clearly 1940s décor. The same guide followed us into this room, enthusiastically explaining the physical placement of everything in the space and frequently using the word "guys" to convey both the excitement and seriousness of the occasion (her phrase repertoire included many statements along the lines of, "guys, look at this"—she simultaneously seemed to want to be our friend and educator). First describing the British presence in Palestine, she then proceeded to show us a picture of the proposed United Nations partition of Israeli/Palestinian land. Such a division of land was not good for the Jews, she said, because Jews were given the desert South and a small area of the North—and did not have a direct route to Jerusalem should "anything" happen.[9] Instead of simply explaining the declaration ceremony to us, she smacked a gavel on the table to open the ceremony and played actual, recorded portions of Ben Gurion's independence speech (in Hebrew, which most of us did not understand) and the Israeli national anthem. Seated where historically important diplomats and formative Zionists used to sit, we the tourist/audience completed the guide's

9 It was left to our imaginations what "anything" would be.

performance. The story she was telling us was unable to be performed without our presence, and by simply "being there," we were engaging in a ritual that had a strong element of fictive play.

The guide's closing statement was personal, making it one of the most emotionally affecting performances of the entire site visit. Once we had been led from the movie theater to the declaration room, our invitation to participate in the constitution and existence of Israel became increasingly explicit. Whereas in the theater the film could conceivably play without our presence or attention, our existence as an audience was necessary for the recreation of a declaration of independence. After the recreation, however, the guide's topic and approach changed—the enthymematic play frame (which in turn produced the enthymematic narrative), rather importantly, did not. The tourist/audience was still necessary for the production; additionally, we, the audience, were implicated as participants within the performance. Our guide's closing remarks were strong with an air of desperation, laying the foundation for an evolving meta-narrative that featured tourists, individuals devoid of peculiar forms of agency, as the central actors. This narrative involved a struggle for peace and survival—a struggle that required participation in the form of military defense, and constant Jewish presence on land defined as Israeli. Israel wants peace, but it also wants to survive, the guide said; we are not given the privilege or luxury of being able to forget and/or ignore the Holocaust. She underscored her belief that Israel really is struggling to survive—she did not truly understand this struggle until she sent her first child into the army, even though she had served for five years herself. When encouraging our support of Israel, it was not about the *learning*, she emphasized, it was about the *feeling*. One of my interviewees, Sophia, experienced a similar tendency of trip organizers to stress emotion and affect over the act of learning, but it upset her immediately. Her story, which took place at the Western Wall, is recounted below.

> Sophia: And so then, the light is fading, it's 4:30, 5:00 by now, and a bunch… The people I ended up hanging out with, the sort of nerdier crowd, um, were like soon it's gonna be sundown there's gonna be no light, how are we going to take pictures, the light is fading, my camera is crap, it's a crap camera, it's especially crap in the dark so that was annoying, and we just wanted to get to the Wall. And it took forever to get there and before we got there we stopped at this little spot where if you were allowed to look over you could overlook it, um the Wall, but they wouldn't let us look they said don't cross beyond here. Then, they said, everybody close your eyes and put your hand on the shoulder of the person next to you in a single file line. I did not like closing my eyes in the middle of Jerusalem, seemed like a bit of a safety problem…
>
> Lisa: Yeah.
>
> S: I don't know, maybe I'm paranoid, some people say I am, yeah, but I like to know what's happening around me especially in a place like Jerusalem. Made us

close our eyes, got in trouble if you opened your eyes, and then what they ended up doing was leading us on this trust walk through these steps that lead towards the Western Wall. So we had to have our eyes closed, so we missed the whole thing going up to the Wall…

L: Oh wow.

S: And at this point me and a couple people were really pissed cuz they were manipulating our experience, like forcing us to have it in this way, they said trust us, it'll be so much better if you see the Wall like this it's the best way to do it. Like, this is…I felt like the whole time we were in Jerusalem they were really making light of the religious aspect of it which is something I'll talk about too. To me, I felt like this was a pilgrimage to go to Jerusalem, you know that's a place that if my family hasn't gone there they would die to go there…Jews for thousands of years have wanted to go to Jerusalem and so many haven't, and we're there and I wanted to have this deep experience and it was…I was prevented because they weren't facilitating it and if I was on my own, I would have been fine doing that myself, right? I could have had it. And this stupid trust walk just ruined it and I was almost in tears it was awful. So finally, you could open your eyes and you could see the Wall. Well, BFD, I could have had a great moment seeing the Wall if I had been allowed to do it without them forcing us [to play] this preschool game. Sorry, it really pisses me off. So then we're standing there and I'm thinking to myself, even though I've read some stuff, I don't really have an understanding of what the Wall is. And that I wanted some more. So I turned to the Israeli guide and I asked her if she could explain some more of the historical significance of where the Wall is in relation to where the temple was, the different parts, what am I looking, you know I feel really ignorant, cuz I don't know, and she looks at me and she goes, puts her hand on my shoulder and goes [Sophia], you know, I guess I'll say it to the whole group but I can say it to you right now if you need, and I said oh no, I don't need it right now, just at some point…[she said] but you know, it's better sometimes just to take in the moment and not worry about the details. It's the moment that really matters, just try to think about that.

Sophia's story is important for several reasons. First, she recounts enthymematic manipulation in one of its more severe forms. Based upon her narrative, it seemed quite difficult to behave in any way other than what the guides were instructing. Sophia's group was not being asked to complete a communicative event, they were being forced to complete it. The anguish she and some of her fellow travelers felt was either not apparent to the group leaders or not important, perhaps a combination of both. In some ways more disturbingly, however, Sophia's Israeli guide subverted knowledge in favor of an uncritical, unreflexive brand of emotion. This subversion in and of itself is not novel, but in this case it is especially noteworthy because Birthright markets itself as both a social and educational experience. The purpose of site visits is often to learn; while it is one thing not to focus upon historical context in Jerusalem, it is quite another to refuse to provide information when asked. In her pre-trip interview, Jordyn, a graduate student, said:

> So it seemed like you would definitely learn a lot more than 'here's the Western Wall,' that there'd definitely be a lot more sociological content that you could either, you know, ask about or maybe not ask about but…I don't know.

When I asked Jordyn what interested her most about the Birthright program, she quickly spoke about the opportunity of having soldiers join the trip. She thought the inclusion of soldiers represented an effort to more fully incorporate tourists into Israeli society—and on many trips, this is likely the case. Famously, military service is mandatory in Israel. As Birthright participants are between the ages of 18 and 26, the vast majority of their Israeli peers (in terms of age) were either currently serving in the military or had recently completed their duties. The placement of soldiers on trips also drew attention to the identity of Israel as a nation-state, rather than simply a tourist location or ethnic enclave.

As it was on Sophia's trip, participants engaged in a kind of responsive numbing. Though she did not feel comfortable or safe, she had to put aside personal feelings in order to be an effective group member. On my trip (and it sounds like on Sophia's as well), disobedient tourists were reprimanded. Although participants could have continued to misbehave in the face of discipline, no one ever did. The threat of being sent back to the States loomed quietly. We had signed forms saying we would behave appropriately, but as it turned out, there was plenty of room for leeway. At times our guides behaved as poorly as we did. Israel's military culture, on occasion, caused tourists on my trip to discipline our leaders.

The Mighty and the Just: Ubiquitous Military Presence

> The violence that is inherent in space enters into conflict with knowledge, which is equally inherent in that space. Power–which is to say violence–divides, then keeps what it has divided in a state of separation; inversely, it reunites–yet keeps whatever it wants in a state of confusion. (Lefebvre, 1991, p. 358)

The trope of necessary and continuous national defense began as early as this first site visit. This story of justified military strength and dominance was developed over the course of the next ten days and, as explained below, became progressively more participatory. A parallel of participation existed between Israeli citizens and Birthright tourists—the Israeli army required participation from every young adult citizen, and our tour likewise called for uniform and total investment. In both cases, involuntary embodied responsibility for collective performance helped establish community, and hence reify statehood. In our tour group, required narrative co-construction by guides and tourists helped legitimize our (American) outsider

Figure 6.1 Birthright Tourists Photograph an Israeli Tank

presence in Israel, making the state feel more like it "belonged" to us—and that we belonged there. When confronted with military equipment such as tanks and guns while on tour, for example, we were not directed away from it, but rather encouraged to examine and touch it (see also Figure 6.2).

Since Israel is not a large country, one is able to travel from one end to the other in a relatively short amount of time—usually about four to five hours, depending upon the speed of your vehicle. Occasionally, we traveled much of the country after sundown when our itinerary demanded we switch cities. One such night, we went from Jerusalem to Tiberias.[10] Our guides told us jokes and stories intermittently as we dozed, woke, and snacked. Our armed guard and medic, Areli, instructed us to look outside our windows at a wall that closed off a Palestinian city. Areli told us he used to guard one of the pillboxes. Perhaps sensing our uncertainty, he explained that he watched Palestinians through bulletproof windows to make sure they were not engaging in terrorist activities. It is highly doubtful that if terrorist activities were to occur in the walled off areas, they would happen right in front of a bulletproof, soldier-laden window, but messages of power and control were as deeply embedded in the wall as the pillboxes themselves.

10 Tiberias, located in the North of Israel, had the most Arab influence/presence of any city we visited; consequently, many of our confrontations with individuals who did not want large groups of touring Jews in Israel occurred on this leg of the journey.

Our third Israeli guide, Hagai, chimed in that we were actually driving by many Arab villages; he stood at the front of the bus and said something akin to, "please be careful, in the event of an attack, these are emergency exits…" and some people chuckled quietly. He went on to briefly poke fun at our "precarious" location, impersonating a flight attendant all the while. Shortly thereafter, our guide Zivah took the microphone. I was relieved we did not have to listen to Hagai anymore, until Zivah started the next joke. It went something like this:

> An Arab, an American, and a Jew are on a plane. They each have to throw something off the plane because the plane is too heavy. The American throws a suitcase full of money because he has lots of that at home. The Arab throws a container of oil because he has plenty of that in his country. The Jew takes the Arab and throws him off the plane, because we have lots of those in our country.

The crowd was not impressed. Very few people laughed, some booed, and Zivah quickly back-pedaled. She wanted us to relax. It's not racist, it's just a joke, she said.

Another troubling occurrence, basic training, occurred on the seventh day of our ten-day tour, the day after eight Israeli soldiers joined our trip. The soldiers were explained more as social members of the group and less as protection; one former soldier (Areli), who had been with us the whole time, carried a gun; those

Figure 6.2 A Tour Bus and Tank at the Israel-Lebanon Border

who joined us on day six did not. The event occurred at night, after a day of touring Yad Vashem[11] and Mount Herzl. The group was collectively exhausted. After six nights of sleeping very little—programming often went well into the night, and it was socially difficult to go to bed right away once the schedule finally took a night's hiatus—we were looking forward to arriving in our new hotel in Tiberias and taking the evening off. Once our bus pulled into the hotel, however, we were told that we would eat a late dinner and meet in the lobby at 10pm for an activity. Wear tennis shoes and comfortable clothes, the guides said. We groaned.

Attendance at all events was mandatory, and this late night bout of mystery exercise was no exception. Several exhausted participants attempted to hide in their hotel rooms, but the guides and soldiers went upstairs to pound on their doors until all participants came downstairs. There was really no other place to hide; the hotel lobby was small and we were occupying it, and it was strictly against the rules to leave the hotel. Once the entire group was assembled in a small room off the lobby, our eight previously kind soldier-peers came in and started screaming. We were going to learn what it was like to be in basic training in the army, they yelled. Arguably, "basic training" was the case in which we had the most communicative limitations placed upon us, yet we also took center stage as performers. The line between performer and audience became especially blurry; as a performer and tourist, however, it was uncomfortable to be forced to violate my own value system (a value system that keeps me out of the army). Screaming at us continually, the soldiers put black marks under our eyes (presumably to keep "sun" out, despite it being 10 o'clock at night and us being inside) and told us to line up in rows of three. "Shut up! Shut up!" the soldiers yelled over and over. Our group was generally quite talkative, but the soldiers wasted no time listening for our voices. After teaching us different formations in which to stand and challenging us to change formations in fifteen seconds or less, our leaders taught us how to fall on the ground and shimmy across the floor, as if we were in battle and needed to get from Point A to Point B without being seen by an enemy. Our tour group of forty was divided into two teams of twenty, and we were told to individually crawl across the floor for the length of the room. We went one by one, and the team who took the shortest amount of time to collectively cross the floor won.

There were several performative discomforts associated with the "mock basic training," and they were often created by the deliberate and disturbing confusion of force and play. In other words, play was intentionally conflated with something that is the antithesis of play—war. I categorize this experience as a form of play because it was clearly make-believe; we were not training to be in the military, and our relationship to the soldiers had already been established as a social, playful one. This idea provides a contrast to Bateson's (2000) theory of play in which he states, "It appears…that play is a phenomenon in which the actions of 'play' are related to, or denote, other actions of 'not play'…It appears, therefore, that the evolution of play may have been an important step in the evolution of communication"

11 Yad Vashem is the Israeli Holocaust Museum located in Jerusalem.

(Bateson, 2000, p. 181). Problematically, however, the distinction between "play" and "not play" became almost irrelevant in this case; we were forced to "play" about very real preparations for combat, and the presence of the currently enlisted soldiers confronted our hypothetical situations with the reality of the soldiers' lives. The "works of affecting presence" (Armstrong, 1981, p. 6) in this case were the soldiers themselves, and a mutually constituted performance was used—by the guides and soldiers—to begin closing the gap of embodied experience between Israeli soldiers and American participants. What is again at issue here is the presence of force (on the part of the guides and soldiers) and lack of consent (on the part of the tourists). Strategic limitation in performance became especially crude and pronounced because we-the-audience were not given the option of not participating in our own transformations. The shift from tourist to soldier-in-training was jarring yet required. Perhaps unwittingly, each individual tourist pressured his or her peers to participate, just by taking part in the performance her/himself. In this sense, an uncomfortable reification of community ensured that we surrendered a degree of our autonomy and agency to the group leaders, allowing them to dictate what we did with our bodies simply by barking an order.

Conclusion

When discussing Birthright Israel, terms like "complicated" and "complex" do not fully account for the events that transpire on the tour. This trip—like many others—is messy. It represents an instance of tourism, but cannot be theorized in light of tourism studies alone; it is persuasive, but cannot be analyzed in terms of rhetorical studies alone; it is performative, but cannot be deconstructed in solely that context. The tour presents several physical sites of contestation, as well as represents an intellectually interdisciplinary site ripe with analytic potential. In terms of geography in particular, however, I aim to develop one final, specific component of my argument.

Through the use of interrelated methods of persuasion, the embodied version of Birthright Israel—and hence its organizers—turn the spatial into the moral. As evidenced by this work, Birthright Israel is not just about the State of Israel. Much to my shock at the time, at its most poignant, the tour was about life and death. Questions of land ownership imply life; at least in many cases in Israel, perhaps due to the small size of the state, to own land is to inhabit it. To inhabit land, or rather to live there, presupposes the right to live—and often to maintain a certain quality of life as well. Most palpably in the enthymematic ways Birthright creates and encourages life valuations and socially enforced relativism, the right to live becomes something that is engineered, manipulated, and justified in peculiar ways.

By infusing Israeli land with the emotionality that comes with thinking deeply about who has the right to live and visit there, trip organizers privilege nationalist feelings at the expense of a critical emphasis upon the content and style of political messages. In other words, an organizational focus on tourists often takes the place of an analytic, reflexive awareness of state metanarratives. Enthymematic tourist experiences assure that State metanarratives are not only perpetuated, but that important aspects of those stories are spoken by people who have yet to be fully persuaded—the tourists. It becomes increasingly difficult for tourists to refuse enthymematic messages because to do so, they would have to deny at least part of a story in which they played a significant role. It is here, in the deep enthymematic structure of the Birthright Israel experience that I found many of the answers to my research questions. The enthymeme binds ever-evolving methods, modes, and effects of persuasion—all the while encouraging the adoption of persuasive, pro-Israel messages. The rhetorical maneuvers identified above operate enthymematically, but the very situation of being a tourist calls for necessary rhetorical participation.

Although there are many ways that questions of space and place are transformed into moral arguments, it is important to note the presence of continuous conflations—conflations of the social and political, social and religious, and religious and political, to name a few. While it is unproductive and likely impossible to separate the aforementioned concepts into mutually exclusive spheres, it is essential that the conflations themselves be regarded critically. In the Birthright case specifically, elements of "the social" are masqueraded as political or religious, and the opposite also occurs. Motivation becomes central to an understanding of intentional and unintentional muddlings of spheres. It is here that perhaps we are best suited to question whether these conflations are meant to raise consciousness, benefit particular groups of individuals, or justify violence.

Chapter 7
Marketing “Danishness”

Introduction

There are thousands of possible sites to see in Copenhagen, Denmark, from the mundane to the iconic and everything in between. One such site might be marketed as follows:

> Historic 9 Svånemosegårdsvej, Frederiksberg, is an old asylum reputed to have a ghost that appears only to children. Today, however, it is comprised of fashionable flats. The slaughterhouse that once sat out back has now been turned into an architect’s office. The garden adjacent to the building houses roses that are over 100 years old.

Yet, while the story of this historic building may generate interest, it and thousands of other sites with similarly fascinating stories are not among the 47 places selected for inclusion in the *Copenhagen Guidebook* (2009), stacks of which line the hallways at Kastrup, Copenhagen’s international airport. This raises questions regarding the politics of tourism site selection processes. The sites listed in the city’s official guidebook are those sites that Wonderful Copenhagen, the official tourist bureau for the city, feels are most emblematic of the city—or, rather, are those that best support the place myth the city and its residents have constructed. Since Copenhagen is the capital of Denmark, these sites might also be considered emblematic of Denmark as a whole. To tour Copenhagen is to glimpse what it means to be a Copenhagener and to be Danish. The two terms are not synonymous; rather, the former is a subset of the latter. Thus, tourism is partially about recognizing and declaring identity—about experiencing one’s own national identity in contrast to an “other” who is being consciously observed.

Identity is a dialectical concept; in constructing the self, we necessarily construct an “other”—one cannot be a member of a group without at the same time excluding others. For example, I cannot proclaim my Danishness without immediately defining those who are not Danish. Yet, identity is also a dynamic concept because its boundaries tend to be unstable. This instability leads to continual redefinition of identity. In Hegelian dialectics, membership in a group and/or exclusion from it lead to a new synthetic membership, whereas in more recent formulations of the dialectic, synthesis remains incomplete, leading to a persistent gap that prevents convergence (see Žižek, 1989). That is to say, in more recent formulations, Danishness and non-Danishness do not lead to a new stable form of Danishness; rather, what is Danish and what is not is constantly in flux.

This gap, this lack of synthesis, importantly does two things. First, it creates an opening, allowing identities to remain fuzzy and therefore personally articulated (Okolie, 2003). Second, the possibilities, raised by an incomplete synthesis of group membership (and thereby self-identity), for intervention and subversion means that identity is often a site of intense political struggle. To utter the phrase, "I am Danish" not only defines those who are not Danish, but simultaneously denotes a willingness to abide by the socially agreed upon (but ever changing) norms, rituals, laws, and material practices of Danishness, and it thus circumscribes citizens. It signals a power relation involving Danes and non-Danes, implicitly and explicitly creating a difference that can be used to distinguish a level of more privileged individuals—Danes in this case—and less privileged non-Danes. However, a difficulty arises when marketing Danishness. As noted in Chapter 4, the norms, rituals, and material practices that constitute identity are internalized. As such, enticing "others" to visit a place means that those norms, rituals and material practices must have some allure, or at the very least be packaged in some way as to have some allure. They must at least hint at universality. There also exists a fundamental semiotic problem, as noted in Chapter 3. Because in tourism promotion imagery is central, that imagery must translate between those creating the images and those consuming them.

How such imagery should translate to the target audience raises the issue of who comprises that target audience. In Denmark, as is the case in most places, fellow Danes are the largest group of tourists. Still, there remains a sizable international tourism flow into Denmark, and these tourists must also be addressed in tourism promotion literature. From the point of view of geography, such a situation raises important questions about the excitability of tourists inside versus outside of the Danish culture and landscape (Cosgrove, 1985). Those who are insiders will likely have considerable background in Danish history, art, and architecture, and will be familiar with the concepts that inform Danish society. Drawing on Chapter 4, one might say Danish citizens are interpellated by Danish society and are versed in its ideographs (see below). Certainly, one can say that as Danish subjects, insiders have a different cache of collateral information available to them than those visiting from elsewhere (see Chapter 3). Conversely, international tourists are often less familiar with Danish society, its history, art, and architecture, and they are neither fully interpellated by Danish society nor are they thoroughly versed in its ideographs. They have a cache of collateral information available to them, but to what extent it may help them interpret what they encounter while on tour is individually contingent. By way of example, suppose a promotional video contains a picture of Rosenborg Castle. To a Dane, this likely would be instantly recognizable and might trigger a series of thoughts about Denmark's longest reigning king, Christian IV, who had the castle built. For an international tourist, the same image might simply yield the response, "nice castle." Logically, then, more explanation must occur in promotions aimed at international tourists. Surprisingly, that is not always the case, as we see below.

In this chapter, we use Denmark as a case study to explore how marketing is intertwined with ideology and semiotics. In teasing out the intricacies of identity politics in the marketing of Danishness, we draw upon two examples. First, we utilize a video advertisement by the Danish tourism organization VisitDenmark. A discussion of the advertisement is useful because it allows us to highlight, in more or less concrete terminology, why a theoretical approach that considers ideology and semiotics as performative matters in a very "real-world" way. Second, we explore the notion of the tourist on-site and the ways in which interpretative materials frame it. Such a consideration is important because, as we have argued in earlier chapters, the tourism site is the primary place of performance. Here, we illustrate these points using various tourism brochures from tourism sites in Denmark.

The Promotional Video

The choice of a tourism advertisement from Denmark is not happenstance, but rather derives from three practical considerations. First, one of the authors of this book has traveled to this country semiannually since 1995, and therefore, is rather familiar with it.[1] Second, Denmark is a small nation whose size makes placing tourism promotion within the larger context of identity promotion easier (see Light, 2001). Third, the Danish economy is export-led, with approximately 85 percent of what is produced in the country exported annually (Knudsen and Kotlen, 2006); thus it has sophisticated international marketing expertise. Yet, somewhat paradoxically, Denmark's tourism promotion efforts have not always been successful (Ooi, 2002; and Copenhagen Post, 2009).

The advertisement chosen here is one produced by VisitDenmark before 2005. As a relatively old advertisement, it is not representative of VisitDenmark's current brand identity and it is used here for illustrative purposes only. The advertisement is built around six key words: design, talented, coziness ("hygge" in Danish), unpretentious, oasis, and free. The advertisement intersperses these key words with scenes from the country as a musical soundtrack plays. Other than the soundtrack, there is no narration. Table 7.1 on p. 114 provides a listing of the themes and scenes from the advertisement along with the duration of each theme.

Let us begin with the naïve level—the level of the potential tourist largely unfamiliar with Denmark. In order to demonstrate this (un)familiarity, we draw on a technique used by Morgan and Pritchard (2005), wherein one author familiar with the country conducts a focus group using the other authors as subjects. This method was employed because the various authors have different exposures to Denmark, which could be utilized in the analysis of the video. Jillian Rickly-Boyd and Michelle Metro-Roland have never been in Denmark, Lisa Braverman

1 Daniel C. Knudsen first traveled to Denmark as a Fulbright scholar in 1995. He has returned there for research most summers since.

spent four months there on overseas study, and Daniel Knudsen lived in Copenhagen from February to July 1995 and has since returned semiannually for research and travel. The focus group method proceeded by having the authors first view the film, and then discuss its meaning and effectiveness in terms of tourism promotion. Specifically, the video was shown three times. First, it was viewed without interruption. Next, it was shown pausing at each theme/scene for comment. Lastly, it was viewed again without interruption. Commentary about the advertisement was recorded on a form that sought authors' thoughts about connections between the scenes and Denmark, along with any specific comments about that connection.

Using this technique, the first recognizable images are those in the "design" theme (Figure 7.1). Here, the assembled images were largely considered emblematic of Denmark and raised thoughts of "arts and crafts movement, northern European modern architecture, livable space, coastal setting," "the north European tradition of bicycling," and "many islands." However, only furniture was noted as distinctly Danish.

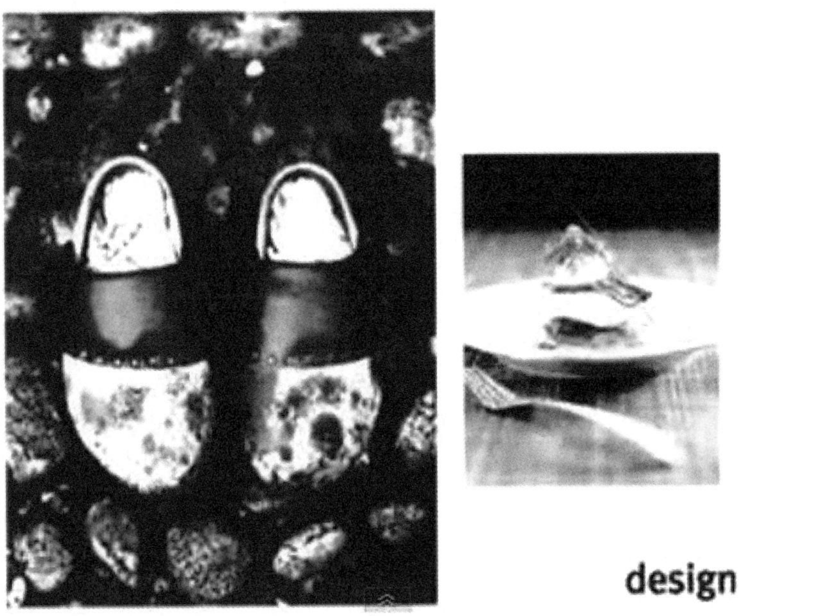

Figure 7.1 "Design"*

* The photographs from this video are used with the permission of VisitDenmark. VisitDenmark wishes to note that this video no longer reflects its current marketing practices.

Under the theme "talented," (Figure 7.2) most scenes were not recognized as emblematic of Denmark. Again, only furniture was considered distinctively Danish or even broadly Scandinavian. One author questioned the degree to which a windmill "implies talent."

Figure 7.2 "Talented"

In the theme of "coziness" (Figure 7.3), many images were again tied to Europe or even northern Europe, but only the images of pastries and flags in the window of a kitchen were cited as distinctively Danish. One author noted that "old homes in the middle of nowhere make me feel anxious, not cozy." Within the theme "unpretentious," (Figure 7.4) only the scenes of the beach, vegetable stand, café, and city market were deemed to fall into that category. And while most images seemed European, none particularly signified Denmark. Indeed, noted one author, the men were wearing "Russian-style hats" and the café was "1950s American."

Regarding the theme "oasis," (Figure 7.5) again few elements conjured Danishness. Among those that did, associations such as "nature, old country with a past, coastal setting, natural setting, history and heritage, and rural/urban divide" surfaced, but as one author pointed out, "the tree could be anywhere, the canoes could be in Michigan, the skating could be in England and the shopping mall in Chicago." In the final theme, "free," (Figure 7.6) notions of "coastal" and "natural" setting predominated. One author noted that the scene of two men together implied that Denmark was "gay-friendly," while another noted that the beach scenes

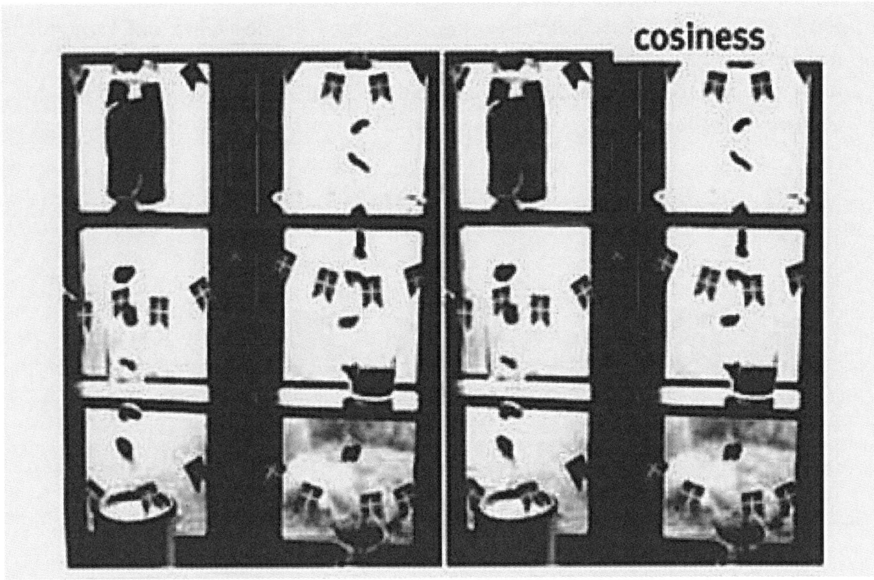

Figure 7.3 “Coziness”

Figure 7.4 “Unpretentious”

Figure 7.5 "Oasis"

Figure 7.6 "Free"

looked cold, implying that this was "not Southern California." A final comment one author noted was, "I recall a few images that stuck with me after watching, in particular towards the middle there is an image of a large window decorated with the Dannebrog (the Danish flag)." Two out of three viewers also explicitly commented that they did not like the music.

Table 7.1 Themes in VisitDenmark Advertisement (Elapsed Time: 3 minutes, 49 Seconds)

Segment	Time elapsed	Description
1: Opening	6 seconds	VisitDenmark seal and slogan
2: Strawberries with cream	10 seconds	Overlaid with six key words used in advertisement
3: Design	33 seconds	Pottery, food, shoes, sculpture, furniture, glass, architecture (phrase "cyber space" appears with "cyber" crossed out), bicycles, long boats, fixed-link bridge to Sweden (phrase "missing link" appears with "missing" crossed out)
4: Talented	15 seconds	Painting, opera, designer clothing, windmills, furniture, food, contemporary music
5: Coziness	35 seconds	Conversation, pubs, old homes, tea time, strawberries with cream, family meals (phrase "bed time" appears with "bed" crossed out), small Danish flags strung together in a window, jazz sessions (phrase "plenary session" appears with "plenary" crossed out), camping, Saint Han's day celebration on the beach, downtown Copenhagen at twilight, pastries, a child eating lunch
6: Unpretentious	31 seconds	A car on the beach with the occupants having a picnic, a child with a hot beverage in a cup, two men chatting holding bicycles, a bar scene, pølse wrapped in newspaper, a countryside vegetable stand (phrase "fast food" appears with "fast" crossed out), ice skating, two men talking at a country fair, a kro, topless sunbathing at Den Konlige Gardens, the inside of a café, a city market (phrase "super market" appears with "super" crossed out)

Table 7.1 Themes in VisitDenmark Advertisement (Elapsed Time: 3 minutes 49 Seconds) *concluded*

Segment	Time elapsed	Description
7: Oasis	39 seconds	Heath, tea time, canoes on a lake, a tree (phrase "traffic light" appears with "traffic" crossed out), a dolmen, the skating rink at Tivoli, a meal at a restaurant (phrase "time table" appears with "time" crossed out), a church interior, an old half-timbered home (phrase "office building" appears with "office" crossed out), a shopping mall, a light house with ship in the background, a meadow, Frederiksborg Castle in the fog, Danish countryside with mustard in bloom, a church steeple
8: Free	49 seconds	Ocean with seagulls, a sailboat under sail, sailboards (phrase "board meeting" appears with "board" crossed out), children running, surf, people wading in the surf, a lake with a wharf (phrase "psycho therapy" appears with "psycho" crossed out), candles, men hugging, a roller coaster, afternoon coffee, a beach with a prone lone person (phrase "corporate life" appears with "corporate" crossed out), child with flowers, flowers, beach and ocean at twilight
9: Closing	5 seconds	Seal and web address

Now let us reread the video from the perspective of the author who is most familiar with Denmark. The opening scene of strawberries and cream is doubly significant (Figure 7.7). First, the combination of the two establishes a palette of red and white—the Danish national colors. Second, strawberries are arguably the most emblematic fruit or vegetable of the Danish summer, the height of the tourist season. Further, overlaying the image of strawberries with cream are the six themes of the advertisement, thereby introducing these tourism slogans to the viewer. The first theme is "design." Illustrations (scenes) of the dominance of Danish design rightfully include pottery (although the iconic Royal Danish porcelain is absent), food, furniture, architecture, Viking ships (which remain some of the fastest ships ever created when sailing with the wind), and bridges (in this case the "fixed-link" bridge to Sweden). Claims of design excellence in shoes, bicycles, glass, or sculpture are dubious at best and the items shown in the video are not representative of the activity in these design fields where Danes have excelled. For example, current styles of shoes that originate in Copenhagen are not shown, nor are the Christiania bicycles, Bornholm glass, or Thorvaldsen's

Figure 7.7 Strawberries and Cream, with Danish Themes Overlaid

sculpture, for which Denmark is most renowned. The second theme is "talented." Here, illustrations include a modern windmill, food, and furniture, all of which are internationally associated with Denmark. The remaining scenes illustrate items in which the Danes themselves take great pride, but for which there is little if any international recognition. While given time in the advertisement, few members of the general populace outside of Denmark could name a Danish painter, opera singer, or contemporary musician.

The third theme, "coziness," while a central tenet of Danish life, is culturally specific. It is the experience of long evenings of good conversation, comfort food, and good drink either outdoors in summer in the long-lingering Danish twilight, or indoors bathed in the light of a multitude of tea candles in other parts of the year. It can also be translated into English as quaintness, and it has strong overtones of the pastoral idyll. For a typical Dane, all the scenes provided in the video as "cozy" are indeed cozy, except perhaps for jazz sessions and camping. Indeed, the highest compliment paid to someone's hospitality is to pronounce time spent with them as "cozy": "Takk vor det. Det var så hyggelig!" (Thanks for it (the evening). It was so cozy!). Not surprisingly, the advertisement features several scenes involving gatherings of friends and family.

The fourth theme is "unpretentious." Here, certain scenes are unambiguous, such as the car on the beach with picnickers, two men chatting, a woman sunbathing (possibly topless) in the Kongen's Have (The King's Garden) in Copenhagen, or the hustle and bustle of a city market. The roadside vegetable stand, the kro (a Danish country inn), and farmers conversing at a county fair are also unpretentious, but could as easily been categorized as "cozy." Still, others are ambiguous: the café scene, a child holding a hot beverage, or ice skating. The theme "oasis" is equally vexed. Scenes of a lake, trees, a church interior, a church steeple, homes, heath lands, farm fields, meadows, restaurants, home interiors (at tea time), and lighthouses are clearly oases. However, demarcating ice skating, dolmens (Neolithic graves), shopping malls, and castles as oases seems a stretch.

The final theme in the advertisement is "free." Here, it is difficult to contest any of the scenes provided—birds, oceans, roller coasters, and children running all evoke freedom. Children with flowers may or may not be free, but they are certainly nice enough. However, only those intimately familiar with Denmark are most likely to understand the connection between tea candles and freedom (they are used each June to mark the liberation of Denmark from Nazi rule). Thus, we see that while some scenes may resonate with their themes for Danes, and even for some international tourists, not all do. For example, even a person who has spent considerable time in the country would have difficulty connecting shopping malls or Neolithic gravestones to an "oasis."

In Chapter 4, we discussed the role that aesthetics plays in tourism promotion and suggested that effective advertisements seek to touch upon all three aesthetic registers—the beautiful, the pastoral, and the sublime. We identified the beautiful with certain features in the built environment (see also Knudsen, Rickly-Boyd, and Greer, 2014), but it is also worth noting that the human body itself may be categorized in this register. We connected the pastoral to a nostalgia for the past that particularly casts rural life as simpler and slower than urban modernity despite the predominance of industrialized agriculture (see Knudsen and Greer, 2011). We connected the sublime to nature, and particularly to those aspects of nature that, while part of the national territory, cannot be fully controlled (see Knudsen and Greer, 2008). If we now consider the use of aesthetic registers in the promotional video, then Table 7.1 indicates that, of 73 scenes, 24 are associated with the beautiful, 33 are associated with the pastoral, and 16 are associated with the sublime. One may map the aesthetic registers fairly neatly onto the promotional video by segment, with the first two substantive segments mainly reflecting the beautiful, the next three the pastoral, and the last a mixture of pastoral and sublime (see Table 7.2). This suggests that the advertisement's creators are using a relatively balanced aesthetic approach with regard to the beautiful and the pastoral—one that stresses modernity, heritage, and tradition. The inclusion of the sublime is illustrative of Denmark's close relationship with the sea. Thus, the use of this combination of aesthetic registers aids in the universality of the advertisement; yet, the specifics that are meant to signify Denmark fail to hit their mark.

In sum, much of the advertisement is obscure enough that those watching it who are not Danish or those without considerable previous experience in the country will simply not understand its symbolism. Among those towards whom the advertisement is targeted, the video will likely fail to create the desire to visit the country, or worse, it will create a desire to visit Europe, but not Denmark

Table 7.2 Segments in VisitDenmark Advertisement (according to aesthetic register)

Segment	Time elapsed	Aesthetic Register
1: VisitDenmark seal and slogan	6 seconds	N/A
2: Strawberries with cream with 6 theme words	10 seconds	N/A
3: Design	33 seconds	Beautiful
4: Talented	15 seconds	Beautiful
5: Coziness	35 seconds	Pastoral
6: Unpretentious	31 seconds	Pastoral
7: Oasis	39 seconds	Pastoral
8: Free	49 seconds	Mixed Pastoral and Sublime
9: Seal and web address	5 seconds	N/A

in particular. Indeed, an empirical test along these lines (see Knudsen and Gray, 2010), indicated that those viewing the advertisement were no more likely to want to visit Denmark than the average population. Conversely, one of the things most notable about Denmark is that it has much of what Europe has to offer (even "Alps"), packed into a very small space. If one combines this attribute with the principles of Peircean semiotics (see Chapter 3)—that tourists' interpretations are, in the first instance, related only to their available collateral information—then an effective advertisement might proclaim:

> Cities, castles, countryside, coastlines…
>
> This is Europe!
>
> This is Denmark!

Such an advertising slogan would draw on more universal themes of "Europe" but nest them within a specific context—Denmark. Of course, this would also require adequate imagery to accompany such a suggestion to potential tourism audiences. This slogan may be effective by making clear to tourists that much of what Europe has to offer can be experienced in a single visit to Denmark, as Denmark has major cities, castles, countryside, and coastlines all within a few hours of each other.

From a rhetorical/persuasive standpoint, this case study offers an illuminating perspective on McGee's (1980) notion of the ideograph. Tourism marketing functions persuasively, as it encourages cross-cultural publics to visit specific locations. In the case of the promotional video, potential tourists are offered slices of Danish ideology: the comforts of coziness, the relief of freedom, and the privilege of talent—among other things. McGee (1980) contends that slogan-like, "God terms" link rhetoric with ideology. Such terms are vague yet all-encompassing, often reifying a culture's foundational values—or at least pretending to. For example, in the United States, common ideographs are "freedom," "liberty," and "equality" (Condit and Lucaites, 1993). The Danish tourism video described here features ideographs prominently, importantly calling attention to ideological modes of cultural reproduction. These cultural slogans, however, are sold in tourism marketing. Tourism marketing is thus able to make ideographs more explicit than they might be in non-tourism situations. Yet, while they may be more explicit, if they do not relate to potential tourists' collateral information, or worse signify other destinations that are not the object of the advertisement, they fail to effectively market place. Culturally specific modes of tacit persuasion are made increasingly visible through both tourism ideology and advertising. Coziness, for example, is not just a practice of comfort. It communicates central Danish values to international tourists, but also strengthens those values among Danes. Such persuasion is made manifest not simply in video advertisements, but also in on-site brochures.

On-Site Brochures

While the role of the tourism promotional video is to create allure leading to a desire to visit Denmark (or any specific place in a different instance, for that matter), the role of on-site brochures is quite different—brochures are designed to mediate tourism performance at the site itself, and as such they play a crucial role in "truth-marking" (MacCannell, 1976). To effectively play this role, on-site brochures must function in concert with collateral information (see Chapter 3). More precisely, to the degree that collateral information, the cache of previous experience each tourist has, leads to inaccurate interpretation and action, on-site brochures are likely to prove helpful in mitigating this confusion. It should also be noted, then, that while on-site materials may be helpful in alleviating confusion, they are ideological representations that frame, and indeed are intended to limit, tourists' interpretations. The information provided to visitors on-site is carefully chosen to communicate particular narratives. Yet, sites receive a variety of tourists, domestic to international, each relying on differing caches of collateral information. In the example used at the beginning of this chapter, the collateral information of Danes is likely to be more useful in the interpretation of a Danish site than, say, the collateral information of an American. Thus, while it might be rather obvious that the international tourist will often need more detailed information than a domestic tourist, with regard to the actual materials presented to tourists this is rarely the case. In the analysis of on-site interpretative brochures that follows, we observe that the multilingual materials offered to international tourists at sites in Denmark contained either the same amount or less information than those in Danish. Further, this information was then observed to contain a number of instances of slippage and misperception which would likely result in inaccurate or precarious interpretations.

In order to begin to explore how sites are described differently across multiple languages, we undertook a simple content analysis of 25 multilingual tourism documents from various tourism sites in Denmark. These documents included text in Danish, English, and German. The analysis involved two steps. First, the length of text in Danish was recorded, then the length of text provided in English, and then the length of text provided in German was recorded. These were then compared. Second, the original Danish and German texts were translated into English and the translated Danish and German versions (and original English texts) were examined semiotically to ascertain what qualitative differences existed between texts.

Of the 25 documents analyzed, 20 provided approximately equal amounts of information in the three languages, while five provided less information in English and German than in Danish. Of those providing less information in English and German than Danish, an average of 5.15 times more text was provided in Danish than in English and 7.84 times more text was provided in Danish than in German. In short, compression was severe. By way of example, the Danish (Jensenius, n.d.):

> A large and grand city plan took form under Frederik V and architect Nicoli Eigtved. St. Anne's plaza, Bredegade, Amaliegade and Frederiksgade was expanded to the current appearance from where they had previously met open fields and an elegant quarter emerged: Frederiksstaden with Amalienborg, four rococo palaces, originally occupied by four nobles families Moltke, Brockdorff, Levetau and Løvenskold, who was unable to finish his mansion and sold it to the Countess Schack, hence its name of Schackenborg. But only after the Christianborg fire of 1794 was Amalienborg the royal residence.

becomes in English:

> [King Frederik V] put the Danish architect Nicoli Eigtved in charge of the planning of the Frederikstown, a grand and fashionable quarter where a great number of imposing mansions were to be built, including what later became the royal residence: Amalienborg.

This shortening omits much of the geographical context for the new city quarter, as well as the historical context for both the building of this section of the city of Copenhagen and for the Amalienborg Palaces. This reduction is a crucial lapse in that much about Fredericksstaden cannot be adequately cognized by tourists not already familiar with its story who are given the shorter description above. The shorter description, for example, is inadequate to explain the (very noticeable) uniform window and cornice heights of this quarter of the city, why there are four Amalienborg palaces and not one single large palace, or why and when the royal family came to occupy this set of buildings.

While documents with shortened descriptions of tourism sites contained significant omissions, translations of documents with approximately equal-length descriptions in Danish, English, and German often omitted cultural details as well. For example, the Danish version:

> On weekends the tour ends with coffee and the delightful Princess Marie layer cake in the private (ground floor) rooms

is translated in the English materials as:

> "Café in the private rooms during the weekend"

and in the German materials as:

> "In the private rooms is the Weekend Café" (Nordfyns Kommune, 2008).

The reference to a layer cake, perhaps the traditional layer cake constructed of successive layers of white cake and whipped cream topped with strawberries and Danish flags, and the explicit reference to coffee is edited out of this passage. Edited in, by either design or simple slippage of language is the word "café" which, to most Americans, is much more inclusive than simply cake and coffee. In the German version of the text, we are told that the café has the particular name "The Weekend Café." One hardly needs to be reminded that weekends are associated with play and relaxation to appreciate how drastically the meaning of the Danish text differs from the German text. Thus, the name "Weekend Café" could mean either that the eatery is only open on the weekend, or that one feels like one is enjoying a carefree weekend when one is at the eating establishment. We suggest, in line with our conceptual framework, that perhaps the more accommodating description would be to have the English and German versions read, "On weekends the tour ends with coffee and Denmark's favorite layer cake in the private (ground floor) rooms," while in Danish, "On weekends the tour ends with layer cake and coffee in the private rooms" is likely sufficient, as most Danes are familiar with the nationally consistent "layer cake." It remains curious that information that could encourage increased cultural competence is repeatedly cut from non-Danish language brochures.

Among texts of approximately equal length, one final slippage can be noted. Consider for example the sentence, "In the northern part of the municipality, you could visit the Tarup-Davinde *gravel pit*, where you can enjoy a delightful landscape with fishing lakes, long barrows, stone exhibition and much, much more" (Midtfyns Kommune, 2008, italics added). Here, there is no difference in translation across the three languages. However, there potentially exist considerable differences in the perceived aesthetic value of gravel pits across the experiences of Danish, American (and other English-speaking), and German tourists. It is this last type of disconnect between text and experience that is the most difficult to overcome, since misperceptions are rooted not in the tourism site or in the textual information about the tourism site, but in the differing collateral information of the tourists themselves. Particularly careful attention should be paid to brochure translations in the tourism site literature. For example, the previous description could become, in the English translation: "In the northern part of the municipality, you might visit the Tarup-Davinde *quarry*, where you can enjoy a delightful landscape with fishing lakes, long barrows, stone exhibition and much, much more." The simple changing of the word "gravel pit" to "quarry" opens up greater possibilities for interpretation and imagination of a pleasing landscape.

Examining on-site brochures within the framework of tourists who are insiders (Danes) and outsiders (international tourists), we have found that, in multilingual on-site brochures, shortening and rewording often occur. Yet, even when this is not the case, the representamen and interpretant (see Chapter 3) rarely come together in similar ways across languages. Indeed, we argue that among on-site brochures we sampled, many are constructed in a way opposite of what relevant semiotic theory suggests would be best. On-site brochures for international tourists would help close the information gap if they were longer, reworded precisely so as to attend to cultural nuance, and sensitive to the ways in which words, even when translated with precision, may have undesirable effects from the perspective of marketing.

Conclusion

This chapter has considered tourism promotion and on-site brochures using ideas drawn from Chapters 3 and 4 to problematize tourism marketing and site/experience interaction. In so doing, an analysis of a Danish tourism promotional video and a content and semiotic analysis of on-site interpretative materials collected at tourism sites in Denmark have been conducted. In the case of the promotional video, the disconnect between the potential audience and those creating the promotional video leads to an unclear and likely ineffective advertisement. Indeed, analysis of the video revealed a number of instances in which the significance of "Danish" imagery was misinterpreted. In the case of the on-site materials, our analysis identifies three areas of concern that might be categorized as slashing, slippage, and misperception. The first, which occurred in a small minority of the texts analyzed, involved the severe reduction of content in translation. The second, slippage, described a language shift in translation so that messages became other than they were intended. The last, misperception, is the most difficult to avoid, and arises simply due to the differing interpretations of the same word across societal contexts.

While this discussion has been largely illustrative, important conclusions can be drawn with respect to the marketing of destinations. First, domestic tourists and international tourists understand tourism sites with reference to substantially different caches of collateral information in part because each is interpellated as citizen/subjects into their society (see Chapters 3 and 4). Second, because domestic and international tourists rely on differing caches of collateral information, they must be courted in different ways and provided different bodies of knowledge at tourism sites (Pritchard and Morgan, 2001). Third, our conceptual framework indicates that international tourists need more information about sites than domestic tourists since international tourists are more often unaware of specific sites' contexts. Yet, content analysis of on-site brochures indicates that sensitive, concerted delivery of site information is often not being carried out in practice.

That tourism promotion often misses its mark and that on-site brochures frequently tell less than the full story is perhaps not surprising. In tourism promotion and on-site mediation, notions of identity often come to the fore, as those creating the promotional and on-site materials draw from their own sets of collateral information—individual and collective. So while the collateral information individuals creating and promoting brochures utilize is likely taken as given or commonsensical, it is indeed ideological. At the same time, those outside the destination culture are at a distinct disadvantage in comprehending the full import of promotional videos and on-site brochures precisely because they typically are not party to the ideological associations of the imagery employed in marketing videos or brochures. In other words, tourism promotion and on-site mediation cannot help but be ideological, but that is also the greatest failing of those practices. In helping define the identity of those at the destination, promoters and tourism bureaus necessarily construct the international tourist as "other."

Chapter 8
Touring Florence

Introduction

In this chapter, we expand upon the points made in Chapter 2 regarding tourism as a secular ritual. We are careful in the framework of this book to draw the distinction between ritual as propounded by MacCannell and Graburn's use of the term. The theory of ritual delineated by Graburn is a better fit for the case studies in this book precisely because it contends that the ritual of tourism ends not at the site, but upon returning home. Graburn's work, in other words, takes account of the interrelationships between a tourism site and a site of "home" in productive ways, and in so doing, also makes room for the rituals of sightseeing to be considered within this larger construct. This chapter focuses on notions of communitas as they play out in two distinct destinations, Venice and Florence, thus exemplifying two particularly different roots of the concept. Venice represents, in this autoethnographic account, a more frequently visited space in which fields of care have begun to develop, thus generating a feeling of community, whereas Florence highlights the often instantaneous nature of communitas.

Through the investigation of a personal travel experience of one of the authors and his spouse,[1] we demonstrate how theoretical notions of ritual, liminality, and communitas are intertwined. The trip to Florence, Italy by way of Venice, we suggest, is generally not unlike the travel experiences of most tourists to Florence or any other unfamiliar space—which is precisely the point. A similar narrative could be constructed about virtually any trip to almost anywhere. The result would be a narrative that is structurally similar, but specialized and personal in its details and impressions. Thus, we may say that tourism is a secular ritual (Graburn, 2010) and that this is demonstrated by a simple retracing of the steps involved in the trip itself.

Pre-Travel Rituals, or Stuck in Limbo…

Preparations for the trip to Florence began more than six months in advance with the purchase of airline tickets. Searching and comparing prices, routes, and dates, we settled on flights from Indianapolis to Atlanta and from there to Venice, with

1 This chapter describes a trip to Venice and Florence, Italy by Daniel C. Knudsen and his wife, Linda, in June 2013. As with Chapter 6, the authors have chosen to retain the personal "I" in some instances here.

our return flights retracing the same path. Flights secured, we then concentrated on choosing accommodations and scheduling round trip transportation between Venice and Florence. Booking a hotel stay in Venice was relatively straightforward, as we had been there before and had a clear preference for a certain hotel. That was not the case in Florence, where several evenings were spent searching websites of prospective hotels before finally settling on a place near the railway station, but not far removed from the center of the city and its sites. We were promised a beautiful view. Shopping for flights and choosing hotels for one's stay are all part of the pre-travel rituals of tourism identified by Graburn (2010) and discussed in chapter 2 of this volume. These rituals engage the imagination and as such, begin the ritual process of tourism.

The purchase of airfare, train tickets, and accommodations was only the beginning of the larger ritual of travel to Florence. We needed new suitcases. Of course we had suitcases, but the ones we had were too large and indistinguishable from others. Considering our previous travels to Europe, we knew there would sizeable distances that must be walked with suitcases in tow, and we opted for smaller, more durable and portable pieces. We also wanted our luggage to be more recognizable on the airport conveyer belt. Looking at luggage options, we saw a sea of black and tan bags promising to get lost among others of their ilk, so we chose a set that caught our eyes—metallic silver with just a hint of blue. Next on the pre-trip agenda: suitable walking shoes and clothing that was both cool and stylish, as we would be there in summer and the Italians are widely known for dressing well. Again, we owned plenty of clothes, but not quite the right ones for Italy, we felt, and our walking shoes were "walked out" and no longer up to several hours per day on granite cobblestones. We wished to "fit in" as best we could. As our departure date grew closer, we began loading up on small, airport security-compliant travel toiletries, books to read on the plane about Florence's history and sites to see, and made arrangements for our pets. All of these are rituals in and of themselves, which in turn are preparatory for the trip. In readying ourselves for travel, we were preparing to step into a liminal period, to be away. Shopping trips represented not only opportunities to acquire new belongings, but to fortify ourselves for an experience that was yet to be determined. Even in a very faint sense, the suitcases, shoes, and clothes we purchased for Florence—most of which we retained even after our trip—hold the aura of having been bought for Florence. Far from making the Florence trip unique, however, this phenomenon simply mirrored the memories attached to items we had bought for trips past.

Departure day came. We went through our list of pre-departure rituals: lights off, stove off, garbage out, dishwasher open, refrigerator closed, air conditioning adjusted so as not to come on until the house reached 82° Fahrenheit (28° Celsius), doors and windows closed and locked, passports, tickets, suitcases, carry-on bags, cash, debit cards, credit cards, and a computer with which to write, post to social media, and Skype were all packed. We drove to the Indianapolis airport and parked in the long-term lot. Thunderstorms were predicted for Atlanta, so we crossed our fingers. We rode transit in from the long-term lot silently, checked in, and turned over our suitcases. Then we headed to security and the post-2001 ritual that travelers must now perform marks the beginning of a liminal space—ticket, passport, shoes off, belt off, computer out, pockets empty, bags into the machine, body into the scanner with hands above your head, meet on the other side and redress.[2] Per usual, there were few objections from fellow passengers about having their bodies and belongings searched. Always a rushed, awkward experience, we made it across the security threshold with only minor irritations and dishevelment. These security rituals accompany the crossing of a threshold and signify, perhaps, more saliently than other pre-trip rituals, the entrance to the liminality of travel. Traversing the security checkpoint of the airport, the traveler enters a non-place (Auge, 1995). It is a space designed to be in-between places. No one belongs here; no one is at home in this space. It is passed through.

Once at our gate, we sat down to wait and watched as the plane was delayed again and yet again. Eventually, the aircraft arrived and we boarded our flight to Atlanta. Our time in the air was short in the scheme of things, only an hour and a half. By the time we had arrived, the thunder showers had moved on, as had our flight to Venice, and so upon arrival, we were shunted from one customer service line to another before being given a discount voucher for a night's stay in Atlanta. We made the best of it and had a nice meal at the surprisingly good hotel restaurant, treating ourselves to the half-gallon margarita specials.

At the 1:00 pm checkout time, we returned to the airport and had six hours to kill before our flight boarded. It is an old saying that all airports look the same, and that is certainly true of Atlanta's airport. It is among the busiest airports in the world. Despite the fact that one can purchase a series of "Atlanta" and "Georgia" themed items in the gift shops, it looks like most others—only bigger. Nondescript signs mark each gate, wide hallways with moving sidewalks accommodate thousands of bustling passengers, and small shops and restaurants pepper the enormous space. Fluorescent lighting illuminates the terminal, though large windows allowed us to see planes arriving and departing. We made our way through security again (we had no bags to check as they had been checked through to our destination the day before, and we already had re-issued tickets from the night before). Re-entering the non-place of the airport terminal, we went to find lunch. We ate our sandwiches more or less in silence, anticipating the next stage of liminality, and then found our gate. It was several more hours, yet, until we were scheduled to board.

In reading these pre-travel rituals, one is, we would imagine, able to identify with this narrative. These rituals are among the most common across travel narratives, for those who utilize this medium of transit. That is precisely the power of the tourism ritual. We individually and collectively perform quite similar rituals, thus illustrating the socially integrating power of ritual. Almost no one

---

2 This further exemplifies the ever-changing nature of rituals. We may have been among the last to experience this post-2001 ritual, as by the fall of 2013 passengers were no longer always required to remove shoes and jackets, and the use of body scanners became tenuous.

bypasses the security checkpoints, for example; celebrity, doctor, or sales clerk, you must follow the proper procedure. Indeed, the ability to bypass this ritual requires a series of other ritual performances, including priority check-in and pre-approval by the (U.S., in this case) Transportation Security Administration. Thus, it is the tourist performance of place that allows for agency and meaning-making, personally as well as collectively. Tourist complicity in the security checkpoint reifies the checkpoint as a valid construct; individual attempts to quicken the security process provide the structure or illusion of agency.

Arrival and Acclimating to Our (Tourism) Place

Travel is the crux of the tourism ritual. It is the point at which one quite literally moves between two worlds. Entry into an airport signals the beginning of a transition across space and through time that has a tourism site at the other end of the journey. Several days or weeks later, one generally returns from that tourism site to one's home in a similar transition. Yet, this transition is also transformational in the sense that one is moving from worker to tourist and thence from tourist to worker. But even as it is transitional and transformational, it is nevertheless marked by class distinction. In first class sit those who travel drinking complimentary cocktails, who have multicourse meals and who can actually lie almost completely horizontal to sleep. Because we had missed the flight the day before, we found ourselves in this class for the first time in our lives. This aspect of the liminality of travel had a special distinctiveness, the aura of which remains vivid in our memories, particularly because we found ourselves again in economy class on the return flight. In economy class, one pays for cocktails, has less luxuriant meals and must sleep sitting up. Regardless of class, transatlantic flight itself has its rituals: One hour from departure drinks are served, a half hour later food is served, an hour later the plane's passengers are "put to bed," an hour and a half before landing passengers are awakened, and an hour before, they are fed.

We awoke bleary eyed an hour and a half out of Venice as we crossed the Alps. Arrival in Venice is always a bit otherworldly (Davis and Marvin, 2004), even when one is coming from someplace else in Europe. This other-worldliness is highlighted when one is arriving by air after a long flight. The airport itself is of a distinctly late 1950s style, and the boat ride across the lagoon to the city itself, whether by waterbus or the much more expensive but considerably faster water taxi, signals that you are somewhere extraordinary. The city itself floats just inches above the Adriatic and all transportation within the city is either on foot or by boat.

This was our second time in Venice, and while that city will not be a focus of this chapter, it is important to say because we have been there before, it is no longer a strictly tourist space for us, but rather a place in Tuanian sense (Tuan, 1974 [1996]). When we have stayed in Venice, each trip has lasted at least a week. Because of our repeated visits, we have fields of care in Venice; we have a neighborhood that is "ours"—Arsenale. Within Arsenale, we are particularly fond of one hotel. There are three restaurants we frequent in which the servers know us and annually say, "Back again this year? It is so good to see you!" Hosts sometimes give us a seat when there is a line or give us the best seat when there are openings. Importantly, many of the locals, intentionally and partially forgetting we are tourists, complain to us about "the tourists" who come to their shop or restaurant wanting this or that. They introduce their spouses and children to us when their families stop by. And we have little rituals we practice: lunch in Murano, the trip to the top of the campanile at St. Giogio Maggiore, breakfast in the hotel dining room, spritzes and sandwiches at the little café across from the entrance to the Arsenale, and drinking wine in the backyard garden of the hotel in the late afternoon before heading off to dinner along the lagoon, a canal, or on a nearby neighborhood piazza. These may be rituals that other tourists perform in Venice as well, but they are perhaps not the norm. On the ground in tourism places, performances vary according to motivation and agenda. One finds points of overlap with other tourists, but also the space to perform your own, individual place.

We have, as a result, moved along the spectrum of tourism from first-time visitor to multiple-time visitor with respect to Venice and it is a qualitatively different kind of tourism experience for us. We deeply care about "our" neighborhood and the few people we know there, and we follow with interest news about the city and about our favorite parts of it, even checking the social media updates of the restaurants we most frequent there from time to time.

On this particular trip, it came time to depart Venice for Florence. To get from one city to the other, we took the high-speed train. Like the transatlantic flight, the trip from Venice to Florence involved both transition and transformation. It involved transition insofar as we moved from a city balanced inches above the sea to one very firmly on the ground, and from a place we knew to space we did not. It was also transformational in that we relocated along the tourist spectrum from returning visitor to first-time tourist. As such, we were filled with anxieties—will we be able to find our way to the hotel, will we like the hotel, where will we eat, how long are the lines for key sites, is the city safe, and so forth.

Approaching Florence

The high-speed Ferrari train (yes, it is red) between Venice and Florence is, like an airliner, tubular in shape. The experiential distinction between airliner and train is further blurred by the last and fastest part of the journey, when the train rushes at 300 km/hr through a series of rounded tunnels before coasting into Florence. There is a distinct pattern, a ritual, to arrival when one is a tourist: one walks to the main exit and gets one's bearings, then walks, takes a cab, or hops on another train in the direction of the hotel—hoping to find either it or somewhere close by where one can obtain directions. After our third attempt at directions, each stop closer to our destination than the one before, we found the hotel. It was not exactly where we expected it to be. We are still at a loss as to how the hotel got the publicity photographs overlooking the basilica they had claimed to be adjacent to their property. The hotel was, however, conveniently located near the tourism district of Florence (Figure 8.1).

Our motivation for visiting Florence was not terribly different from that of other tourists, we suspect. We came to see the city that is the birthplace of the Italian Renaissance—the first true flowering of Greco-Roman culture at the end of the Dark Ages that marks the beginning of the modern European period. The tourism district of Florence is remarkable for both its volume of tourists and its compactness. This combination creates a quasi-festival atmosphere, one in which large numbers of tourists are participating in the same activities and normal conventions, and the proscription against talking to strangers breaks down. Such an atmosphere in turn, leads to a high level of interaction among tourists in this space. Similar to Venice, there are a number of "must see" tourism sites; however, unlike Venice, navigation does not pose quite the challenge. Florence's tourism district stretches from the Basilica of Santa Maria Novella to the Pitti Palace to the Basilica of Santa Croce to the Galleria dell'Accademia and then back to the Basilica of Santa Maria Novella.

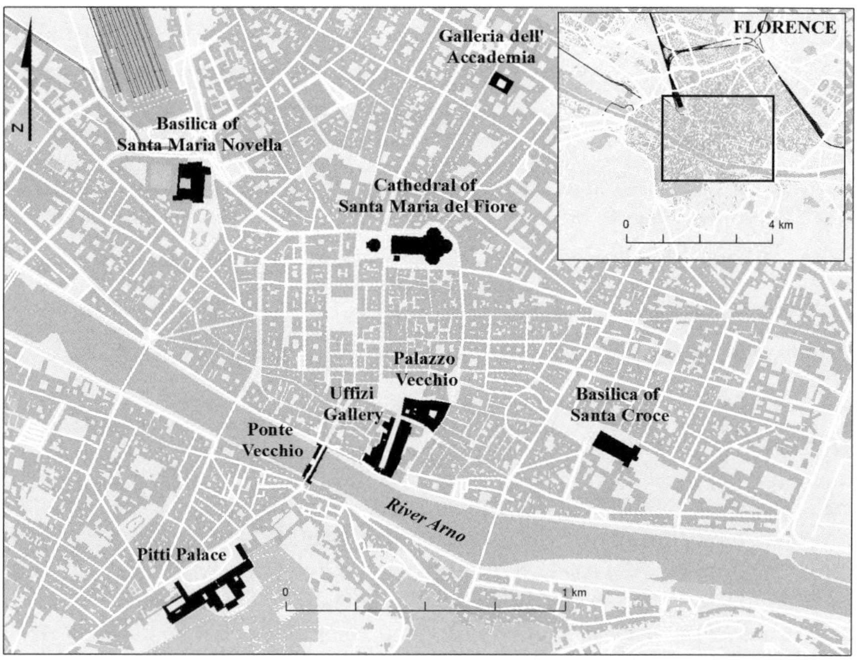

Figure 8.1 Florence, Italy (by David P. Massey)

Within this larger circumference lay the Duomo (Cathedral of Santa Maria del Fiore) with its Baptistery and Campanile, the Palazzo Vecchio, the Uffizi Gallery and the Ponte Vecchio. One can walk the entire circumference of the tourism district in a little more than an hour and point-to-point distances are usually 20 minutes or less. Within the confines of this tourism area, the crush of bodies is considerable. Touring Florence bears strong similarity to secular pilgrimage, partially accounting for the city's immense popularity (Theilman, 1987; and Cohen, 1992). As in many cities, there are "must see" sites. The Accademia and Uffizi Gallery are at the top of most Florence tourists' lists and thus have the longest lines (Figure 8.2). After discovering this bit of information, we asked other tourists we met about the lines and they pointed us to a place where we could purchase tickets in advance, cutting waiting times to half an hour. We toured remaining sites while waiting for admission to the Accademia and the Uffizi. It was a tip we passed on, along

Figure 8.2 Tourists in Florence, Italy

with the location of the advanced sales ticket office, to other tourists we met in Florence. Once inside the Accademia there is considerable jostling to get close to Michelangelo's David; the same is true of the various Botticellis at the Uffizi. Just about everywhere within the tourist zone, tourists stand snapping photographs of one another with a "must see" site in the background, acting out Jenkins' (2003) circle of representation—having seen pictures of Florence, they have now come to see it in person and to collect photographs of themselves to prove to others that they have been there. Through such performances, tourists situate themselves as individuals among a tradition of touring practices, thus enacting MacCannell's ritual of sightseeing.

As discussed in Chapter 4, a ritual is a repetitive, communicative performance that involves actors presenting signs to an audience that already knows the meaning of those signs. Thus, rituals have semiotic content, and indeed, tourism sites are signs (Knudsen and Rickly-Boyd, 2012) that require performance and interpretation by tourists. Most sites in Florence require little of the tourist beyond a general knowledge of Western civilization, art and architectural history, and aesthetics in order to be understood. These sites remain historically significant, however, through their continual ritualization. That tourists *do* tour them, take and share photographs, and share their experiences with others performatively maintains their roles as signs. Indeed, the principal draw for global tourists is that Florence is an important stopping point in history as the center of the Italian Renaissance. The work of Botticelli and Michelangelo exemplify Renaissance concepts of beauty (which owe much to Western classical conceptions of the same). Like the work of Botticelli and Michelangelo, the Duomo and its Baptistery, which share an exterior of white, green, and red marble are things of great beauty, but at a much grander scale. Turning a corner from a narrow alleyway as one makes her or his way toward central Florence, one is struck immediately by the enormity of the Duomo, which is among the largest Christian churches in the world and holds 30,000 worshippers (Figure 8.3).

Two other sites in Florence, the Palazzo Vecchio and the Pitti Palace, require a firmer handle on the politics of Renaissance Italy and thus were more difficult for us to interpret. Particularly the Palazzo Vecchio, which has served as the town hall of Florence for over seven hundred years, requires considerable knowledge of Florentine politics—and this we largely lacked. Thus it was difficult to appreciate exactly that at which we were looking. Both are extremely popular sites in Florence known for their beautiful interiors and this is why we went to see them, yet both also functioned as ritual performances we did not completely understand. We may have arrived at the site and completed, in performance, the ritual act of sightseeing. Yet, our social integration was not complete. We remained alienated from some of the significance of these spaces, not fully immersed in the tradition of Renaissance Italy. Tourism, however, is about experiences beyond historical traditions. It extends to sociality, intersubjectivity and communitas with other travelers.

Figure 8.3 Florentine Architecture: The Duomo

Cafés, Strangers, Communitas

Given the close proximity of the most popular sites in Florence and the large number of tourists there, it is not surprising that tourists encounter one another in the cafés that cluster the squares. There tourists exchange information on sites, the best ways to get tickets, good restaurants, bars, hotels, and so on. It was through this process of encounter and exchange that we learned the best way to get tickets to popular attractions, thus avoiding the long lines. In situations like these, tourists create an instantaneous sense of community. The trigger for the beginning of a conversation is most typically a common language and a question between two tourists. The conversation frequently extends well beyond tourism and includes information that one normally would not share with a stranger—where one is staying, where one is from, why one is in this or that city, and where one has been. The liminality of the tourist experience explains this sharing of information—because the experience itself is removed from everyday life, typical societal rules are eased. Also, because tourists are in a liminal space, the need for information-sharing is greater.

The following exchange overheard at a sidewalk café (Figure 8.4) on the Piazza del Duomo in Florence is illustrative. Sitting adjacent to a man and woman from Australia who were sipping "Sky Spritzes" (an Italian drink made with a licorice-flavored blue aperitif), we were all chatting while waiting for our drinks to arrive. Two young American women arrived on bicycles, parked in the street, and entered the café to sit near the Australian pair. "Where should we eat?" one says to another. "I haven't any idea," the other responds. A moment passes as our drinks arrive, and then a few more minutes elapse in silence as the women take in the Duomo. One of the Australian tourists then leans toward the Americans and

Figure 8.4 Florence Café

points, "There's a nice place there; it's not too pricey and the food is good." Before long, the Australian pair and American women have exchanged a great deal of information. The Americans are new arrivals in Florence, but will be here for the summer and fall studying art history. They are from Chicago, are currently staying in a village just outside of Florence, and use bicycles as their primary mode of transportation into town. The Australians had been in Florence for five days and would leave tomorrow for Venice, where they would join a cruise of the Adriatic. They arrived in Florence from Rome and before that, London. Additionally, they live on Australia's Gold Coast and are retired teachers. We leave to have dinner as they order another round of drinks and continue their conversation. They have formed a bond of sorts, an instantaneous community. Despite there being only a few people in this particular community, kinship inspired by the liminality of tourism differs from casual interactions outside of it. This liminal space highlights individual ignorance in a way that managing everyday life does not. In navigating the uncertain spaces of tourism ritual, advice from a stranger has a greater chance of radically altering one's day.

The notion that an instantaneous community can be formed from such a quick interaction might seem like an overstatement, but we argue it is not. The social nature of humans and our propensity to form communities is, of course, well-known. The sense of community or communitas that the formation of bonds engenders has traditionally been explained as having a basis in shared attributes or properties, be they real or imagined (Anderson, 1983; and Smith, 1991), or in fields of care (Tuan, 1996 [1974]). In those times when we experience communitas, we are enveloped in a "homogeneous totality," an "ethically superior human condition where equality, humility, and unselfishness spontaneously prevail" (Sandall, 2011, p. 483). In the case of tourism, the notion of communitas is often attributed to the fact that we are all tourists, and in that sense, we are all performing the same (post) modern ritual (Turner, 1969; Bauman, 1996; and Di Giovine, 2011). However, this attribution tells only part of the story. Communitas in the case of tourism also is dependent on the breaking down of social barriers that occurs within liminal situations (Sharpe, 2005). Personal details become, in a sense, less personal in a liminal space.

Even this explanation, however, fails to account for the example above. What is missing from the explanation is a sense of obligation of one tourist for another, as captured in the action of the Australians' intervention in the conversation between the two American women. An alternate explanation of the sort of instantaneous community one finds in tourism can be found within the framework discussed by Esposito (2010), who theorizes community as a gift, a benefit that entails an unspoken obligation that, while perhaps never called in, remains in force. Within this framework, what binds tourists together is our need to belong coupled with our obligation to render assistance to others in the tourism community—precisely because as members of that community, we are obligated to assist. It is this latter obligation that likely led the Australian woman to say, pointing out as she did, "There's a nice place there; it's not too pricey and the food is good." This same impulse operated when we allowed two New Zealanders to sit at our table at an outdoor café near Basilica Santa Maria Nouvella because, unlike others in the café, we did not smoke.

Despite these shared moments, even on our last evening in Florence, having completed the tourism rituals of the city, we did not feel close to the city. We had approached the city, but it was not "ours" in the sense that Arsenale in Venice is "ours." We left behind no field of care. We made no connection; met no one who might say "glad you are back" should we go there again. We met other tourists and had wonderful moments of conversation, but these were fleeting, no contacts were established, just memories of serendipitous encounters and exchanges. Perhaps it was the presence of so many tourists in such a small space, perhaps it takes a longer stay in the city, perhaps it takes staying elsewhere than in a hotel, we cannot say. While Florence never became "ours," we did feel a sense of having seen its sites. We had performed the rituals of Florence's most popular sites—the Duomo, Baptistery and Campanile of Santa Maria del Fiore, Michelangelo's David, the Palazzo Vecchio, the Ponte Vecchio and the Pitti Palace. Each of these holds an important place in the Italian Renaissance and as such, in Western aesthetics and cultural heritage. We now know what each of these places is, why (at least generally) they are important, and what it is to be in their presence. In that sense, completing this ritual offers the potential for a degree of social integration (MacCannell, 1976) but that integration does not promise the forging of an emotional bond with a specific place. Nevertheless, as Tuan (1974 [1994]) reminds us, place is experienced at multiple scales. So while we may not have established fields of care in Florence, as we have in Venice, it is a public symbol. Further, as Thrift (1999) states, places are passings that haunt. Indeed, in our touristic performance of Florence, some experiences, in the form of memories, remain with us well after the duration of the tour.

Returning Home

From Florence, we returned to spend a few days in Venice. The stop in Venice provided a transition to "home" in the United States, as we felt much more comfortable in Venice than Florence. In Venice, a strike by boatmen led to our final days being spent mostly on foot in the Arsenale neighborhood. As we had no particular agenda and had walked a great deal in Florence, we were not overly disturbed by the turn of events. On the morning of our departure, we splurged and took a water taxi to the airport. The flight home was long, but mercifully uneventful. We arrived home late in the evening and, exhausted, fell into bed. It would be a day or two before we were fully accustomed to being in the United States again.

While the return trip itself is simply a retracing of steps, a return from the liminality that tourism provides, the tourism ritual is normally quite unlike the traditional, sacred ritual (Turner, 1969) in that talking about what happened while away is a crucial part of the secular ritual itself. And while the return home marks a return to most aspects of "normal" life, the sharing of adventures while away through stories and photographs and the gifting of souvenirs to friends and neighbors are part of the post-trip ritual in a similar way to sorting through the mail or unpacking the suitcases.

It is also the case that this post-trip portion of the ritual has been altered by electronic forms of communication. The ability to Skype with friends and loved ones while on tour, and the ability to post to social media or to blog means that those at home can monitor your daily adventures—or even several times a day—and can know what you are doing while on tour. The ability to communicate with home occasionally also affected the liminality of our trip. Especially when Skyping, we were no longer fully away, in a semi-private liminal space. That said, at the end of a period of electronic communication, we were able to slide back into moments of liminality. The ability to instantly communicate effectually re-centers post-trip rituals around that which is not visual—the sounds, smells and tastes of the trip that cannot be communicated (effectively or affectively) over the Internet alone. This re-centering of the post-trip ritual away from the visual, in turn, influences souvenir choices. This explains partly why we increasingly bring food-based souvenirs back from our travels for our friends and loved ones. In so doing, we seek to bring to them a part of our experience that they may share through their senses.

Conclusion

This chapter has worked to illustrate the points made in Chapter 2 about tourism as secular ritual. For us, the theory of ritual delineated by Graburn seems a better fit precisely because it considers that the ritual of tourism ends not at the site, but upon returning home. Neither theory fully considers the notion of communitas, hence we include it here as a central element of the tourism experience. Communitas assists in structuring our time on tour and provides a ready-made group identity, helping to ensure our ability to interact and navigate in strange places. We argue, as did Turner (1969), that communitas is a salient dimension of the tourism experience and one that helps explain tourism's attraction. For it is the liminal aspects of the tour that provide for the rearranging of normal societal order that in turn allows new bonds of community to be formed. These bonds are typically fleeting, manifesting themselves in a pleasant evening of company after a random encounter at a café in Florence, or perhaps even in a simple gesture of help as in pointing the way to a good restaurant, a ticket booth, or a train station.

Additionally, electronic intermediation means the liminality of tourism is becoming more permeable. Life at home increasingly bleeds into life away and life away into life at home. This affects all parts of the secular ritual that is tourism and alters, to some degree, the pre-travel preparations, the trip itself, and the post-travel telling of stories and sharing of photographs. This places, we argue, greater weight on the non-visual elements of travel as these are, importantly, the more difficult aspects to communicate upon return. They remain ours, largely impenetrable by others.

More broadly, situating MacCannell's ritual of sightseeing into Graburn's framework of tourism as a secular ritual also speaks to the ways touristic performances enact places. Because Graburn's tourism begins and ends at "home," the ways in which places are forged through mobility and connectivity are more salient. It is through mobility that Venice and Florence were made accessible. Touristic performances of Venice have resulted in community networks that extend well beyond our "home" to include individuals who live in the city. In so doing, Venice has become a field of care. Touristic performances of Florence included participation in rituals of sightseeing that brought us into contact with socio-cultural sites of significance and facilitated numerous communitas experiences. Yet, for us personally, Florence did not foster a sense of place that then results in a field of care; a sense of distance remained between us and the city. As a place, it continues to haunt—our passing through yielded memories of Florence that remain with us. These memories, however, do not evoke a sense of belonging that encourages us to return any time soon.

Chapter 9

Lifestyle Rock Climbers: Mobile Performances of Home

Introduction

Tourism is premised on a number of dichotomies—home/away, guest/host, and work/leisure, to mention a few. Likewise, many forms of tourism are focused on specific locations and being in place. Thus tourism's definitional boundaries, which delineate time away from home as a central variable, can be problematic when addressing movement-based, extended, or long-term travel. As such, many tourism scholars are beginning to consider tourism as an extension of social, political, and economic practices, rather than a temporary aspect of life that is separated from our everyday lifeworlds (Hannam, 2009, p. 106; see also Pons, 2003). As a result, some advocate for tourism to be understood and examined as a specific process within the wider context of mobilities, thereby situating tourism with other movement-based concepts (Hannam, 2009; see also Chapter 10). Mobilities studies are concerned with the politics of movement, in which movement is considered the "'spatialization of time and the temporalization of space'" (Hannam, 2009, p. 102).

Mobility is just as essential to the human experience of the world as place, argues Cresswell (2006, p. 3). Indeed, the politics of mobility (and immobility) accompany the transformation of Western European society from a feudal system of highly restricted mobility through the rise of capitalism, which brought about the right to individual mobility. Research on lifestyle travel, in particular, highlights the mobilities turn. Lifestyle travelers are individuals who pursue extended periods of leisure travel; it is a preferred lifestyle to which they repeatedly return (Cohen, 2010a, p. 117). This travel mode, thereby, illustrates a de-differentiation of travel and everyday life (Cohen, 2011). While it is now receiving more scholarly attention, this is not a new phenomenon; long-term travelers have also been studied under other names—"lifelong wanderers" (Noy and Cohen, 2005), "contemporary drifters" (Cohen, 2004), and "nomads from affluence" (Cohen, 1973). This research has primarily focused on motivations and experiences of lifestyle mobilities as well as the economics of long-term travel (Riley, 1988; Obenour, 2004; Cohen, 2010a; 2010b; 2011; Filho, 2010; Benson and O'Reilly, 2009; and Benson, 2011). What has been overlooked, however, is the notion of privilege as a precondition to this form of mobility. In contrast to migrant workers whose mobility is driven by economic necessity, the various manifestations of lifestyle mobilities (travelers, migrants) *choose* hypermobile lifestyles. It is a means of enacting an aspect of personal identity. Most have the financial means to prepare and save for long-term travel and the option to return to a sedentary life. They make the choice to give up many material possessions, to transform a vehicle into a mobile abode, and to live minimally. This lifestyle originates in a place of privilege, framed in terms of Western liberal individualism in which one's primary responsibility is to the self. This theorization is not to diminish the experiences or motivations of lifestyle travelers, but to acknowledge the differing preconditions for lifestyle mobilities.

This chapter works in the border zone between tourism and mobilities studies, engaging performance-based theories to consider the case of lifestyle rock climbers. Because these individuals spend months to years to decades in the pursuit of rock climbing, they exemplify the precondition of privilege as well as the shortcomings of both tourism and mobilities literatures to capture the nuances of hypermobile lifestyles. Thinking about space through movement is precisely what performance theories do, argues Thrift (1999). Performance theories have shifted analytical perspectives in human geography, illuminating the embodied production of space, the agency of the non-human, and representations as doings (see Thrift, 1996; 1999; 2000; Anderson and Harrison, 2010; and Macpherson, 2010; see also Chapter 5). In particular, this chapter considers the concept of home, a relatively stable idea used to define tourism, but one that needs rethinking as we approach and cross the boundaries between tourism, travel, and mobilities. What is "home" for the lifestyle traveler or in relation to lifestyle mobilities? How is home performed on the road? What implications does this have for notions of place and for considering tourism places as points of encounter and connection?

Research Design

Within the rock climbing community there exists a number of subcultural identities relating to style of climbing, regional preferences, and degree of dedication to the sport (from leisure to lifestyle to professional climbers). As a subculture, lifestyle climbers exhibit a passionate dedication to the sport by maintaining a minimalist, hypermobile lifestyle. While non-sponsored and not earning an income from their climbing, these climbers give up sedentary residences, living out of vans most commonly, as they travel between climbing destinations and take up temporary and/or internet-based employment along the way. Spending years in the pursuit of rock climbing, lifestyle climbers develop strong community networks, foster independence and self-sufficiency, and in so doing, challenge the dichotomous notions of home/away and work/leisure.

This chapter is a part of a larger research project which investigates lifestyle climber motivations and experiences, travel behavior, and community dynamics. By their very nature, hypermobile communities are constantly in movement, with individuals traveling a variety of circuits and pathways, thus making them rather difficult to study. Therefore, this research focused on one popular climbing destination—Red River Gorge, Kentucky—during its optimal climbing season, September through November, as lifestyle climbers arrived for the season, mooring for a few weeks to months, then departing for their next destination. An ethnographic study of this community resulted in a total of 21 interviews with lifestyle climbers—6 females and 15 males. The age of participants ranged from 22 to 56 years, with the time spent traveling and climbing from six months to 17 years.

In addition, analysis also extended to the popular climbing website, Rockclimbing.com. In an age of increasing globalization, social media and networks can foster communities that span vast distances (Massey, 1994; and Hall, 1995). While climbers do meet and perform group identities on the ground, in specific locations, the website's forums facilitate community development and maintenance. Altheide et al., (2008) suggest that "an ethnographic perspective can be brought to bear on symbolic communication in other than 'physical spaces', including information bases and cyberspace" (p. 135, see also Hine, 2008). Accordingly, discourse analysis of the website's forums provides insight into the social relationships and travel behavior that characterize this sport.

The (Hyper)Mobile Home: Community, Belonging, and Dwelling

The rock climbing community in the U.S. traces its roots to Yosemite and pre-WWII climbing clubs along the West coast, in particular. Following the war, however, the community changed drastically. The 1950s, Taylor (2010) argues, brought the "individualists" and the "experientialists" to the fore. These climbers began venturing onto rock faces alone or in small groups outside the organization of clubs; in so doing they pushed the limits of what was considered climbable. The Beatnik climbers of this generation, in particular, fashioned an entirely new ideology, "one that simultaneously honored tradition and championed a countercultural quest for authentic experience" (Taylor, 2010, p. 131). This group is especially noted for bringing the elements of travel and climbing together, rather than simply focusing on Yosemite or other regional hotspots as many prior had done; the "experientialists" took to the road and followed the climbing season (Mellor, 2001; and Taylor, 2010).

Among the first to turn climbing into a lifestyle was Mark Powell, who began a seasonal rhythm of winter work, while saving money and training in Los Angeles, followed by a move to Yosemite for climbing spring through fall. By the late 1950s, Powell was "simply the best," and by dedicating himself to the pursuit of the experience he changed the sport (Taylor, 2010, p. 134). A fellow climber wrote of the time, "he showed us all that climbing can be a way of life and a basis for a philosophy" (Pratt, 1965, p. 346). Taylor (2010) identifies several commonalities among the Beatnik climbers that made their perspectives on society and the sport far different from their predecessors:

> These communities were filled with loners and outcasts by chance and design. [...] More common were some very well-educated young men rebelling against a world that had groomed them. Nearly all attended college, most had degrees, and a few were in graduate school before choosing climbing instead. As in the past, Beats embraced the sport as a way to construct the self [...] It was a tightly knit yet fluid society, with members coming and going through the year: spring in Yosemite and the Southwest, summer in the Tetons, Rockies, and Alaska, and fall back in the valley. (pp. 138-140)

As this group pushed the limits of their bodies, their sanity, and their substance tolerance, they also pushed the limits of the sport. Moreover, the Beats were the first generation to make a living from climbing, earning money from writing essays and guidebooks, developing new gear, or acting as climbing and wilderness guides in order to support their full time rock climbing; sponsorship opportunities came only to the later generations.

By the late 1960s and into the 1970s, the social scene of the rock climbing world had evolved so that the Beats were the "elders" and they viewed the next generation as overly "arrogant," "vulgar," and lacking commitment (Taylor, 2010). While they saw their own "limit-pushing" as philosophical, they regarded these "dirtbags" as interested only in recreation, over-indulging in everything from poverty to drugs (Childs, 1999; and Taylor, 2010). Yet, it is the "dirtbag" generation that is credited with the discovery of new climbing areas and broadening travel circuits while simultaneously fashioning more specific climbing styles.

The top climbers of the 1980s, then, took advantage of new opportunities that afforded them a much more comfortable lifestyle. The proliferation of climbing gear companies fostered international competitions and sponsorships, along with climbing media. Following this proliferation, some began to allow their likeness to be used in advertising of outdoor gear; others became centerpieces for outdoor recreational photography and videography (Taylor, 2010). Thus arose the figure of the professional rock climber, notably Lynn Hill and Ron Kauk in the early 1990s.

Choosing the term "lifestyle climbers" to describe the full-time, non-professional rock climbers of today is an effort to capture the diversity of manifestations observed, while also relating to the bodies of literature regarding lifestyle mobilities and lifestyle travel. In fact, these climbers use a series of self-identifiers; common to all of them is the notion that this is a lifestyle and that they are not professional rock climbers. "Dirtbags" is the most common subcultural identity, used to connect individuals with a historical lineage and ideology in the rock climbing community. However, some prefer to be identified as "lifers" or "full-timers," as a way to express their dedication to the sport, while others use "vanner" in order to convey their minimalist lifestyle, travel mode and most common abode.

Rock Climbing as a Lifestyle

Lifestyle is understood as,

> a distinctive set of shared patterns of tangible behavior that is organized around a set of coherent interests or social conditions or both, that is explained and justified by a set of related values, attitudes, and orientations and that, under certain conditions, becomes the basis for a separate, common social identity. (Stebbens, 1997, p. 350)

When choosing to climb full-time, these climbers are also choosing a life of travel. Explained by one climber, "there's nowhere that is great 12 months of the year, so if you really want to climb all the time, it's pretty important [to travel]." This results in a hypermobile lifestyle, with only a few weeks to months at any single destination. Thus, lifestyle climbers illustrate the spectrum from lifestyle travel to lifestyle mobility. Whereas those who live this lifestyle for a few years or less are better characterized as lifestyle travelers (Cohen, 2010a; 2010b; 2011), the climbers that maintain their dedication for longer periods of time, a decade or more, can be better understood as engaged in lifestyle mobility (Benson and O'Reilly, 2009; Wilson, 2009; Filho, 2010; and Benson, 2011). The majority of lifestyle climbers fall somewhere between these two types. Common to both mobile lifestyle pursuits is the significance of movement. Mobility and travel are essential and frame these identities.

While constant movement between climbing areas is the primary shared behavior of lifestyle climbing, the lifeways that result at such destinations create a particular set of social conditions. As this climber describes, performing a lifestyle dedicated to rock climbing necessitates certain living conditions that would not be chosen otherwise.

> I think it's sort of a lifestyle. I'm living behind a pizza shop in an RV, but if I didn't climb I would NOT be here. There's no way. I would not live in a car, or RV, or tent, in the backwoods of Kentucky. (Female, mid-20s, lifestyle climber of 2 years)

These shared social conditions, that is, minimalist, mobile lifestyles and mooring at out-of-the-way locations is expressive of the lifestyle climber community. Shared among this community, therefore, is an attitude and set of values for which rock climbing is central and is held in reverence. So while they are all "climbers," generally, their lifestyle dedication informs their travel behavior and thereby their subcultural identity. One climber described what bonds this subculture of the rock climbing community: "respect for other people's experience and for the specialness of what it can be out there sometimes" (Male, late 20s).

Belonging is about *feeling* at home (Yuval-Davis, 2006). Community is a primary sense of belonging (Cohen, 1985). Although belonging is a subjective feeling, it is also socially defined; that is, membership is not sufficiently defined by the individual, but it must be validated by others in the community (Fortier, 1999; and Ralph and Staeheli, 2011). The strongest source of community cohesion among lifestyle climbers is an intense dedication to rock climbing, so much so that it is the practice of climbing that guides life choices and travel behavior. Explained by this climber,

> I'd say a dirtbag is just someone who lets the act of the climbing dictate their life. [...] And most of the dirtbags are out there because climbing means something to them, they believe in it and they believe it's going to do something to them. I'd go a step further and say they don't even believe in it because then that sounds sort of faith-based. They know it, they've seen it make a change in their life and they keep pursuing that. Somebody who is making their decisions based on rock climbing. […] They're not here for the social scene, they're here for the climbing. And all that other stuff exists as a byproduct. (Male, late 20s, lifestyle climber of 10 years)

Because this high degree of reverence is upheld as being of the greatest importance for acceptance as a lifestyle climber, this community exhibits a great internal variety in terms of national and regional origins, political viewpoints, religious affiliations, and education levels. Lifestyle sport subcultures are "fundamentally about 'doing it', about taking part" and participating in the appropriated subcultural spaces (Wheaton, 2004, p.4). A minimalist lifestyle is particularly important in this regard as it extends beyond a financial necessity to a mechanism of community cohesion; it brings climbers to the same campgrounds where they socialize, as well as collectively organize their days of climbing and rest.

The highly exclusive social conditions of this lifestyle pursuit foster strong community bonds, despite the hypermobility of individual lifestyle climbers along various circuits of destinations. Individuals may only see one another for a few weeks out of the year; yet, they seldom experience a sense of distance and separation. In fact, farewell performances are rare.

> I guess, with dirtbags, when you say good-bye to each other, it doesn't really mean anything. […] It ends up not being a really long time until you see each other again anyway because you're all in the same circuit. You're all chasing the good weather or the country where the best, new, interesting climbing over the winter is. Yeah, nobody really, like, dwells on, like, the whole goodbye thing in this community, because everybody's always around. You say goodbye to somebody, they're like I'm sure I'll see you in the next year or so, whatever. (Female, late 20s, lifestyle climber of 1 year)

Belonging emerges out of "entwined social processes of incorporation and exclusion that are partly self-defined, partly other-defined" (Ralph and Staeheli, 2011, p. 523; see also Young, 1990; and Fortier, 1999). For lifestyle climbers, community is about the climbing and the climbing experience. The prioritizing of this practice thus necessitates both spatial proximity and distance. While proximity fosters the development of social bonds, spatial distance is a factor of this lifestyle and community that is accepted as part of what it means to be hypermobile. So while the concept of communitas (Wang, 1999) has been taken up in tourism studies to capture the spontaneous, temporary social bonds of travel (see Chapter 8), it is not adequate to fully explain the sociality and sense of belonging that develops among lifestyle climbers. Travelers may experience communitas, as individuals meet in serendipitous locations and then depart in separate directions unlikely to meet again. But lifestyle climbers, because their travel patterns are informed by rock climbing, are likely to meet again and again over the course of months and years. Further, social networks develop as a result of these encounters so that one may be new to the social "scene" of a climbing area but still have mutual relationships that facilitate inclusion.

Home on the Road

While traditionally home has been considered from a sedentarist, discrete perspective, more recent scholarship suggests this concept is better understood as both an experience of a location and a set of relationships (Ralph and Staeheli, 2011). Home is first and foremost socially defined. As such, Nowicka (2007) observes that for the hypermobile, home is maintained through social networks. Moreover, Germann Molz (2008, p. 137) argues, with increased mobility, "home becomes a signifier not only for the normative stability of a particular place or for the transportable sentiments of comfort, security, familiarity, and control, but also for a way of being and belonging in the world as a whole" (see also Nowicka, 2007).

This is particularly the case for lifestyle climbers. As discussed above, the social relations and lifeways that result from this lifestyle pursuit are essential to notions of community and belonging. While the full-time pursuit of rock climbing necessitates travel, individual levels of mobility vary greatly. Some spend no more than a few weeks in a single destination, whereas others divide the year among only two or three locations. Regardless of the frequency of travel, home becomes mobile (Figure 9.1).

> Home is just wherever I am. I've got my van. I've got my own space here, when I need to just be alone. (Female, mid-30s, lifestyle climber of 5 years)

As many explained, not only is there the physical shaping of an automobile into a mobile abode but also the social construction of space. The space of the van, in this case, becomes home—one's only personal space of comfort, security, familiarity, and control while on the road. The physical shaping and personal enactment of a vehicle into a home space thus suggests a place of privilege from which lifestyle climbers imagine and plan their mobile lifestyles. For some individuals without permanent homes, the automobile is always already a dwelling

Figure 9.1 Mobile Abodes of Lifestyle Rock Climbers

space. For lifestyle climbers, however, much time and effort are invested in the physical transformation of the vehicle so that it may be appropriated as dwelling space. Indeed, it is not uncommon to find climbers who have removed the interiors of their vans (seats and carpeting) to then build in a bed frame, cabinetry, and storage, as well as install extra electrical wiring and outlets.

Yet, home "has more to do with the issues of dwelling, organizing, arranging your life," argues Nowicka (2007, p. 77); it is a strategy for dealing with the constant interplay of absence and presence that comes with hypermobility. This climber hints at the emotional turmoil of mobility versus spatial stability,

> You're constantly moving, constantly, there's no, like, real home except for where you are. And I think that's a real great way to live. No matter what I'm doing, if I'm traveling or if I'm living in one place, home is where I am. But sometimes it gets a little hard on the cold nights and the waning moon; it's tiresome. You start looking for some type of stability. (Female, early 30s, lifestyle climber of 1 year)

For those who have maintained this lifestyle mobility for longer periods of time, the nuances of mobility and home become habitual. In so doing, home comes to mean more than a space, but also relations—human, non-human, and self.

> It's just been within the past year or two that I've realized there's not like geographical areas that are home, so it's been pretty easy to be like home is the

> back of my Toyota. I have everything I need with me and I can bring it wherever I want. And I am home in my body, instead of like, I have to be in a certain place. So traveling has felt pretty normal, pretty mellow. (Female, early 30s, lifestyle climber of 7 years)

Lifestyle climbers make home along the way, rather than a place from which they leave and to which they return. Thus, the daily practices of domesticity and dwelling are more significant to the production of home space than a singular, physical location. This also suggests home is not the closed-off entity it was once considered. While home is defined by restricted access and a sense of security, it is also open and porous, connecting to various networks across time and space and adapted accordingly (Nowicka, 2007; and Massey, 2005).

Mobile Dwelling

Heidegger originally developed the concept of dwelling in relation to *Dasein*, being-in-the-world, and rootedness. Yet, exploring the mechanisms of dwelling as a practice has made it a fruitful addition to tourism and mobilities studies (Ingold, 2000; and Pons, 2003). When Heidegger (1971, p. 142) argued, "building and dwelling are related as end and means [...] building is not merely a means and a way toward dwelling—to build is in itself already to dwell," he was suggesting that "to dwell" has more to do with involvement in the world and less about residing in a particular location (Dreyfus, 1993). When incorporating the concept of dwelling into tourism studies, "the starting point is the recognition that being is always being-in-the-world, that is, a situated and contingent process of engagement" (Pons, 2003, p. 49; see also Dreyfus, 1993). Thus, in the investigation of mobile performances of home the concept of dwelling becomes essential.

Home for the hypermobile is an ongoing process of emplacement in the context of mobility through which sets of human and non-human relationships, along with material objects, are given priority (Nowicka, 2007). Movement is the primary means of being-in-the-world for lifestyle climbers, as rock climbers and as travelers. As such, community networks and familiarity with others' being-in-the-world function as essential components in the performance of mobile dwelling.

> You see the same folks out there doing the same thing. Usually friendly with one another, and that's usually your climbing partners, who you're going to be psyched on. Living the similar lifestyle, anything from helping you fix your brakes [...] to free dinners to rides. (Male, late 20s, lifestyle climber of 10 years)

> Living in a transient way finds only the people that are living the same way as you. They become your family. (Female, early 30s, lifestyle climber of 1 year)

Yet, non-human relationships and material objects also maintain a sense of familiarity across various destinations. Although material possessions are minimal, they are significant. Living in a small space, every object has a purpose—utilitarian or sentimental—and was chosen to be included with these parameters in mind. As one climber notes,

> I go through, like, monthly and yearly editing sessions where I go through everything that I have and just take out what I don't need 'cause space is a premium. So the things that I have in this van…if anyone stole this van, I would be lost. I have, like, the important things that aren't of any importance to anybody else—like my grandma's quilt she made me…this cup…every single thing in this van is, like, super edited down. I've just surrounded myself with the most important, sentimental things. (Male, late 20s, lifestyle climber of 7 years)

The materialities of the hypermobile home help maintain connections to relationships that are spatially distant. Pons argues travelers must be thought of in terms of integrated, complex networks of things, technologies, and places (2003, p. 61; see also Lury, 1997). These hybrid assemblages include intimate interdependences of human and non-human relationships (Pons, 2003). Such non-human relationships even extend, for some, to cohabitation of the home space with a companion animal, usually a dog. While this adds to the familiarity and intimacy of the home space, it also necessitates daily routines. The needs of another being become incorporated into personal, mundane rituals of hygiene, sleep, and meal preparations, as well as the daily practices of rock climbing (the primary motivator of this lifestyle pursuit), which include packing gear and hiking to the cliff line (Figure 9.2). "For world travelers, home comes to be found in a routine set of practices, in a repetition of habitual social interactions," argues Nowicka (2007, p. 72; see also Rapport, 1994). As such, dwelling is fundamentally practical, not cognitive (Pons, 2003, p. 49; and Dreyfus, 1993), as this climber illustrates:

> Actually, when I go home and I go to my parents' house, you know where I go to bed at night? I go to bed in my truck. My parents have like eight bedrooms and I go to bed in my truck. Because it's my space, it's my apartment; it's my home. And I sleep well. I close the doors and I feel most comfortable, I feel most happy living out of my truck. (Male, mid-30s, lifestyle climber of 12 years)

It is important to remember in the production of space, it is the body that is the medium through which we are practically involved in the world (Pons, 2003, p. 55). As Merleau-Ponty and Lefebvre have both argued, the body inhabits space/time but also spatializes, it makes space of and for the body. So while home may

Figure 9.2 Everyday Rituals of Lifestyle Climbers

be a space of comfort and security, limitations on the spatial body can also lead to experiences of confinement, frustration, and discomfort.

> I'm not that old, but I've worked my body really hard for working and climbing, so it's like, for me, getting out of bed and not being able to [fully] stand up in the van to get dressed and my fucking back hurting and my shit everywhere…I guess it's the little things that just whittle away at you over time. (Male, late 30s, lifestyle climber of 6 years)

Likewise, each climber interviewed had a story that expressed the limitations of her or his mobile abode as a dwelling space. Many stories involved harsh weather and extreme heat and cold, and others told of illnesses and injuries that resigned them to the confines of "home." Yet, the home they had made, in these situations, did not offer comfort and security but led them to further assess their lifestyle motivations. Thus, it is worthwhile to note the choice that these lifestyle travelers have made in this pursuit, which continually surfaces as an issue of contemplation.

Rethinking Heidegger's concept of dwelling for increasingly mobile lifestyles means considering dwelling as lived, embodied, situated, and relational practices of home, wherever that may be (Pons, 2003; Ingold, 2000; and Dreyfus, 1993). Moreover, it opens up the idea of home, highlighting that the everyday practices, materialities, and social relations that shape home resonate with the individual body, and at the same time, far beyond its borders (Blunt and Varley, 2004).

Home/Place

Tuan (1975) argues that experience makes place at various scales; the most personal, individual, and intimate place is home (see Chapter 1). Home is for the sustenance of the body, as "the one place in which we can openly and comfortably admit our frailty and our bodily needs" (Tuan, 1975, p. 154). But home is more than the distinct location of our departures and returns, it is made through the practice of dwelling, the relationships that weave through and connect individuals across space and time, and materialities that offer familiarity and comfort. "People dwell in and through being at home and away, through the dialectic of roots and routes" (Nowicka, 2007, p. 72; see also Hall, 1995; and Gustafson, 2001); in so doing home becomes mobile. Considering home as mobile, therefore, also has implications for how we think about place.

While mobility and place were once considered mutually exclusive (Relph, 1976; 1981), the performative turn in geography has shed light on the relational qualities of place (Thrift, 1999; Cresswell, 2002; and Simonsen, 2008). As such, place is now considered to be the intertwining process of roots *and* routes (Hall, 1995; Clifford, 1997; Gustafson, 2001; Massey, 1991; 2005; and Baerenholdt and Granas 2008). Without the weaving together of these relations—absence and presence, emplacement and displacement—"place would simply not 'happen'"(Johannesson and Baerenholdt, 2008, p. 155). Place is not construed out of "nowhere," but as points of encounter, mooring, and connection.

Moving from the intimate, personal scale of home to place for lifestyle climbers means paying more attention to collective travel behavior that brings individuals to similar destinations over time. Lifestyle climbers choose their destinations by individual preferences in climbing style and seasonality, but also personal circumstances. Nevertheless, travel circuits overlap, community is performed, and relationships are fostered and forged in the social production of place. Perhaps more importantly, as *the* motivator of this lifestyle pursuit, climbing places are embodied spaces. They are produced by the act of rock climbing. Climbing destinations would not exist without the act of climbing and the knowledge of these acts being spread through community networks. Crouch (2002, p. 214) argues, "[t]he subject doing tourism makes lay knowledge through a complexity of awareness that is immediate, diffuse, and interactive [...] we 'know' places bodily and through an active intersubjectivity." Indeed, lifestyle climbers' initial motivations to pursue rock climbing full-time necessitate frequent and continuous travel. It is through travel that place is encountered. Yet, place is not in the locality, but in the embodied performances that encourage mooring and the sociality that is born out of shared perspectives on rock climbing. Climbing places are thus points of connection and mooring. In their performances of climbing places, and individual practices of home and dwelling along the way and in-between climbing destinations, lifestyle climbers produce constellations of places. Like home, place is porous; it is enacted through embodied practices as well as sociality that fosters community networks across places.

Conclusion

This chapter has worked to rethink the concept of home in the context of long-term travel, lifestyle travel, and lifestyle mobilities. While the act of tourism is defined by time away from home culminating in a return, it should be asked: What happens to "home" when one travels for extended periods of time or is continuously on the move? Understanding how home can be mobile correlates with practices that also constitute place. Home, argue Ralph and Staeheli (2011, p. 525), "is like an accordion, in that it both stretches to expand outwards to distant and remote places, while also squeezing to embed people in their proximate and immediate locales and social relations."

For the hypermobile, home does not lose its spatiality, rather it becomes a performative, mobile space. As such, home is in the relationships that stretch across destinations—with each new climbing area comes a new set of faces along with familiar ones that thereby maintain a community of climbers living minimally as they dedicate themselves to the sport. It is a community made from the dialectics of proximity and distance. But lifestyle climbers also produce small, intimate spaces out of their means of travel. The practice of dwelling becomes mobile with daily, mundane routines of domesticity, material objects of personal significance, and non-human relationships that are carried out along the way. Home is also, then, a space of and for the body. As Tuan (1975) asserts, home is for the sustenance of the body, and for lifestyle climbers this becomes particularly evident. Time spent in the home space is in traveling between rock climbing destinations, in rest and recovery of the climbing body, and in the nursing of injuries sustained in the act the climbing. Thus, home is at once grounded and uprooted, it is the result of "mobility *and* stasis, displacement *and* placement, as well as roots *and* routes" (Ralph and Staeheli, 2011, p. 519; see also Gustafson, 2001).

In following Tuan's (1975) assertion that space is made meaningful at multiple scales, it is not surprising that it is the embodied practices, human and non-human relationships, and material encounters that also result in place. Place for lifestyle climbers is experienced as points of encounter, connection, mooring, and spatially distant networks of people and practices. Turning to performance-based theories in both human geography and tourism studies thus illuminates the practices that extend beyond destinations to the pathways, connections, and relationships that grow out of travel. And in so doing, such perspectives allow us to uncover the nuances of tourism, travel, and mobilities.

Chapter 10
Conclusion: Expanding Tourism Geographies

On December 29, 2013, singer/songwriter Ani DiFranco announced that she would be canceling the "Righteous Retreat," a workshop she had intended to lead. The retreat was billed as a combination songwriting/jam session/fieldtrip. In addition to offering the public creative time with DiFranco and several other artists from the Righteous Records label, the retreat embraced strong elements of tourism, including a New Orleans day trip. The entire event was canceled, however, after significant public outrage regarding the retreat's venue. DiFranco's retreat was scheduled to be hosted at Louisiana's Nottaway Plantation ("Righteous Retreat Cancelled").

DiFranco, known for music that radically engages issues related to feminism and anti-colonial, anti-racist politics (among other topics), has developed a fan base that is sensitive to such topics—and those fans were far from thrilled about the retreat's scheduled location. Reactions to the event as they impacted DiFranco's persona are surely worthy of study, but our primary concern here is with the animation this case brings to intersections of tourism, performance, and place. In becoming outraged by the staging of a songwriting retreat at a plantation that clearly supported the enslavement of human beings, those who voiced their anger about the event brought notions of place to the fore. The plantation's history became part of its public persona; the site performed simply by existing. Indeed, as we have argued throughout this book, place matters in tourism.

A blogger who credits her/himself as "Riese," in analyzing the Righteous Retreat through the lens of "dark" tourism, argues:

> The horror of a concentration camp is clear-cut, but plantations actively obscure their darkness. Prisons and asylums are not, generally, beautiful buildings, whereas plantations are opulent and majestic by definition. Nobody wants to spend the night at Alcatraz, but Nottoway functions as a hotel and wedding venue. There are no memorial walls listing names of slaves beaten, killed or raped at Nottoway. Whereas [some] sites of mass genocide are clearly advertised as such, plantations are relentlessly dishonest, designed and celebrated to bury a violent legacy. ("Out of Rage")

Under substantial pressure from her fan base, DiFranco decided to cancel the event altogether. She expressed both solidarity with and surprise at her fans, noting she did not anticipate that the choice of Nottoway would elicit such passionate responses. Four days after her cancelation notice, she again posted to her record label website—this time offering a short, heartfelt apology and promising to "dig deeper" ("From Ani").

DiFranco's Righteous Retreat, and her fans' reactions to it, thus exemplifies the relationality of the theoretical perspectives put forward in this book. The landscape of Louisiana's Nottoway Plantation, while a contemporary tourism destination, carries with it a past of slavery and dehumanization. The ideologies that accompanied that time period are made manifest in the present through the architecture of the estate. Times and social circumstances may have changed, but the landscape as palimpsest does not erase what came before. Of course, ideology retains its power through reproduction. So while Nottoway may be read for particular histories, it is also performed in various ways. One might argue that DiFranco's proposed Righteous Retreat could be characterized as a means to perform the space in a different way. Such a performance does not mean ignoring the past, but altering the ways in which we approach a place's history and the ways that history is reproduced through our interactions with place. Framing tourism as performances of places opens up places to new and multiple ideologies that problematize their pasts in constructive ways by keeping them continually present.

Contributions of the Book

Surveying the tourism literature, it has come to our attention that despite increased scholarly attention to tourism, there remains a persistent divide in the ways researchers approach this phenomenon—focusing on either the tourism site or the tourist experience. In writing this book, we have worked not only to strengthen the literature that connects the tourist experience with tourism places, but also to render the connections between ritual, semiotics, ideology, and performance more explicit. Further, we suggest that a geographic approach, one employing a landscape perspective to the performance of place, is particularly well-suited to hold together experience/site investigations.

In Part One, the theoretical frameworks from which tourism can be approached—ritual, semiotics, ideology, and performance—were delineated and expanded upon. By dedicating a chapter to each of these perspectives, we aimed to illustrate the far-reaching implications of each while also illustrating their relations and the necessity of conversations between and among them. We seek to better articulate the tourism experience and tourism site under a more encompassing framework of investigation. In Part Two, we moved to ground tourism performances in places as a means to illustrate the ways in which ritual, semiotics, and ideology work in performative ways and in relation to one another.

Writing a book on tourism theory means considering what has come before, how tourism (and how we think about tourism) has changed and is changing, and the ways we can account for these flows. It does not mean making a final statement on the matter. Indeed, that lack of finality proves to be one of the more frustrating aspects of the process. By the time such a task is completed, the loose ends become all too apparent. We set out to write this text in order to work through one, albeit a considerable, problem in tourism studies at the moment: How do we better stitch together the tourist experience and tourism site under the same framework? We feel we have been able to offer a sound means to do so, and hope this book encourages others to continue to investigate and flesh out this problem. At the end of this task, however, a few areas of interest remain that were not explicitly taken up in this book but, we feel, are directly related.

Some of the earliest theories of tourism identify alienation as a motivating force (see Chapter 2). While engaging with this line of conversation, we have chosen not to delve into the breadth of scholarship that has developed around alienation's dialectic—authenticity. Authenticity is among the most contentious concepts in tourism studies, with debates being waged over its meaning, categorizations, and even usefulness as a concept. While we have not engaged with the concept directly, the relations of tourism motivation, experience, and place performance map onto authenticity in ways similarly outlined in this text. Authenticity, as a multi-faceted concept—objective, constructive, postmodern, existential—does inform tourist motivation; it shapes expectations of place, and thereby, is influential to touristic experience. Authenticity is also used among tourism practitioners and informs the development and marketing of tourism sites. Of course, different and multiple registers of authenticity may be (and frequently are) employed in different circumstances. The concept is pervasive in tourism. Further, it has been suggested that various registers of authenticity are relational in the performance of tourism (see Knudsen and Waade, 2010; and Rickly-Boyd, 2012). Ritual, semiotics, and ideology are essential to understanding the ways authenticity is used by tourists and tourism practitioners.

One of the implicit goals of writing this book, and engaging a geographic perspective on tourism, was to test the boundaries of tourism geographies as they currently stand. While attending to a landscape perspective on tourism places as performances has allowed a richer account for the intricacies that interlace tourist experience and tourism sites, we have also repeatedly found ourselves bumping up against definitional boundaries that have become foundational in tourism studies, notably home/away, work/leisure, self/other, and insider/outsider. As such, we are increasingly finding the mobilities paradigm a worthwhile framework for situating tourism along with other movement-based phenomenon (see also Baerenholdt et al., 2004). Mobility is meaning ascribed to movement, and therefore, mobilities studies works to dismantle categorical boundaries of movement types (tourism, migration, circulation, etc.) so as to foster conversations that articulate, instead, the politics of mobility (and immobility). The cases presented in this text range from those that are more explicitly tourism-based to those bordering on circulation as a means of illustrating the ways various mobilities work together and inform one another and, thus, support the move towards investigations of tourism mobilities.

A Mobilities Perspective

Mobilities studies is a capacious field concerned with a number of movement-based phenomena, from everyday circulation and automobility to transnationalism and diaspora. Sheller and Urry (2006) call attention to the mobilities turn across the social sciences, stating that the mobilities paradigm better frames the ways in which people's daily lives are spatially connected. This includes the politics that drive (and hinder) the movement of people as well as objects, information, and non-human things. As a result, Hannam (2009) suggests tourism would be better approached as a specific process within a wider ontological context of mobilities. Advocating a mobilities turn suggests tourism is more usefully considered in the context of other movement-based concepts. Indeed, such a shift in perspective is leading tourism scholars to theorize the phenomenon less as "an ephemeral aspect of social life that is practiced outside normal, everyday life" but instead as "integral to wider processes of economic and political development and even constitutive of everyday life" (Hannam, 2009, p.106). Tourism, while a liminal space/time, is not completely severed from our daily lives. On tour we continue many of our daily rituals (see Chapter 3), we maintain connections to home and/or work through global telecommunications (see Chapter 8), yet we also confront unfamiliar ideologies (see Chapters 4, 6, and 7). Tourism forces individuals to come to terms with what is familiar and unfamiliar, while incorporating these encounters with place into one's understanding of the world and self (see Chapters 2, 5, and 9). As such, tourism forges connections to and across places.

While traditionally tourism has been distinguished as a unique phenomenon, with its own industry and impacts across economic, political, social, and cultural realms, examining its complexity as an endeavor undertaken by individuals reveals many overlaps with other movement-based activities. Mobilities studies, by taking movement, displacement, and the relationship between mobility and immobility as starting points, sheds light on the processes and practices that underlie tourism. As Hannam (2009) argues, "[n]ot only does a mobilities perspective lead us to discard our usual notions of spatiality and scale, but it also undermines existing linear assumptions about temporality and timing, which often assume that actors are able to do only one thing at a time, and that events follow each other in a linear order" (p. 109; see also Hannam et al., 2006).

Further, many forms of tourism are very much focused on specific locations and being in particular places. Such a perspective considers place as bounded, which, then, quickly becomes problematic given the various ways tourism actually happens. Further, approaching place as insular hints at the ways hypermobility and globalization once induced fears of homogeneity and placelessness. Indeed, as Crouch (2010, p. 10) describes, "[t]he fleeting view from the car window is now familiarly offered as stereotypical of contemporary mobility, detached from other kinds of practice in a way that renders visual cues dominant, landscape passing by and emerging serially as in a movie." Yet, Baerenholdt and Granas (2008) contend that connections to places and senses of belonging can actually become stronger through mobility. Mobility fosters new connections to places that expand one's networks of social relations and senses of place. From a mobilities perspective, place is "throwntogetherness" (Massey, 2005; 1991), place is constituted as nodes in continually expanding and interweaving spatial networks of life. Movement, Hannam explains, "is made up of time and space; more clearly it is the 'spatialization of time and the temporalization of space'" (2009, p. 102). Mobility understood as socialized movement (Cresswell, 1999) complements, then, a "progressive sense of place" as a "constellation of social relations, meetings, and weaving together" (Massey, 1991, p. 28).

Like performance-based theories, mobilities studies concerns the enacting of space and place through the connectivity of a globalizing world. Baerenholdt and Granas (2008, p. 2) explain,

> We are [...] in a state beyond the dichotomy of the good local, so-called 'internal', control versus the bad non-local, 'external' control. Connections and encounters crucial to people's lives are often much more complex and dynamic [...]. Contexts are thus not predetermined at any scalar level, but only emerge with the practices of making and becoming places and mobilities.

Thus, place is informed by practice. As a result, moorings are just as important as mobilities, particularly in understanding "constellations of mobility"—the patterns, representations, and ways of practicing movement that become entangled and informed by particular politics of mobility (Cresswell, 2010). Yet, the new mobilities paradigm does not view "immobilities" as irrelevant; mobilities studies is necessarily about different kinds of potential—not only movement with meaning, but also who and what have the ability to be mobile.

We agree with the point that mobility is "just as central to the human experience of the world, as place" (Cresswell, 2006, p. 3). Indeed, the social acceptance of mobility illustrates the transformation of Western society from a feudal system of low mobility through the rise of capitalism and the right to mobility, accompanied by the modern figure of the tourist. Thus, the time period and style of mobility has held significant social and political meanings throughout time, from the pilgrim to the tramp to the tourist (Cresswell, 2010).

Situating tourism within a mobilities paradigm and attending to the politics of mobility thus better articulates the nuances of newly-identified tourism mobilities. Long-term, multi-sited travel, for example, has long problematized the nuances of tourism and illustrates the potential of a mobilities perspective (see Cohen, 2010a; 2010b; 2011; Benson, 2011; and Benson and O'Reilly, 2009). Lifestyle travelers, as Cohen (2010a) explains, are "individuals for whom extended leisure travel is a preferred lifestyle that they return to repeatedly" (p. 117). These long-term travelers have also been referred to as "lifelong wanderers" (Noy and Cohen, 2005), "contemporary drifters" (Cohen, 2004), and "nomads from affluence" (Cohen, 1973). Such travelers push the conceptual boundaries of the tourism ritual, often not returning home in short order (if at all). "Home" thus becomes a more fluid concept as tourism itself morphs and changes. Further, lifestyle travel is inherently political. It illustrates issues of privilege and the right to mobility.

Medical tourism is also a more recently theorized iteration of tourist mobility that is exemplary of the right to move, economic privilege, and also the politics that drive one to tour. The (in)ability or monetary cost of acquiring particular health services in one's home country drives those of financial means to travel to another country in order to receive potentially life-saving medical services (see Kingsbury et al., 2012). This form of tourism mobility is largely characterized by the travel of those of the Global North to the Global South. As Connell (2006) notes, medical tourism is booming for non-life threatening procedures such as cosmetic surgery. Many hospitals that consciously engage in medical tourism—for example, in India and Thailand—have low infection rates comparable to the most successful hospitals in the countries that supply medical tourists; indeed, according to Connell (2006), medical tourism is well on its way to supplying billions of dollars to destination countries. As with the types of tourism discussed below, medical tourism necessitates innovative performances of self and place. Tourists need not travel to a place simply for the promise of enjoyment and pleasure.

Birth tourism further highlights the blurring of tourism mobilities and migration. In the case of birth tourism, a pregnant mother will travel to another country in order to give birth and secure citizenship for her child. This iteration of tourism mobility thus reverses the flow exemplified by much medical tourism in that many mothers travel from the Global South to the Global North, with the United States being a frequent destination. The U.S. is a particularly popular destination in this regard, as the citizenship requirements are based on place of birth and because the perceptions of the economic opportunities that U.S. citizenship may bring (Ormonde, 2012). Birth tourism thus emphasizes the importance of the concept of potential to mobilities studies. Countries seen as having more potential to provide successful lives to new citizens inspire more movement.

As we indicated in Chapter 5, types of tourism that Bowman and Pezzullo (2009) importantly problematize as bearing the label of being "dark"—disaster, toxic, or death tourisms, for example—are proliferating. The memorial and museum of Auschwitz and Birkenau, according to the location's website, are steadily receiving more visitors each year. Over a million tourists came to Auschwitz and Birkenau in 2006; over 1.4 million visited in 2011 ("Record Number of Visitors…"). Fukushima, Japan, is becoming an up-and-coming hybrid of disaster and toxic tourism (for a thorough development of the concept of the toxic tour, see Pezzullo, 2007). Blending these types of touring with imperatives to remember tragedy and to strive for the creation of a better world has been the impetus for the conceptualization of a tourist village at Fukushima, reports the *Telegraph,* as of August 2013. Despite the nuclear disaster of 2011, protective hotels, restaurants, and a museum are being planned ("Japan's Tsunami-Hit Fukushima…"). As the same article boasts, "dressed in protective suits and wearing respirators, tourists will be able to take photos of the shattered reactor buildings and the workers who are still trying to render the reactors safe." Such tourism, taking shape in a landscape that has yet to fully assess the impacts of radiation on surrounding life, enacts a complex and confused performance of place. A location like Fukushima is simultaneously safe and not-safe, welcoming while at the same time keeping tourists at a distance.

Each of these examples, and we could go on with more, illustrates the potential of a mobilities perspective for situating tourism among movement-based phenomena, rather than tying it so strictly to leisure and/or recreation. A mobilities perspective allows for assessing the complexities that drive movement, as well as hinder and regulate it. It allows researchers to move past the complications of the definitional boundaries of tourism to interrogate the politics that inform travel and thereby inform performances of places. As such, it complements and extends the performative turn in geography by offering greater consideration to embodiment, agency, representations as doings, and connectivity of places.

Bibliography

Adloff, F., 2006. Beyond interests and norms: Toward a theory of gift-giving and reciprocity in modern societies. *Constellations,* 13(3). pp. 407-427.

Agricultural Marketing Resource Center, (nd). Available at: http://www.agmrc.org/.

Agnew, J., 1987. *The United States in the world economy.* Cambridge, UK: Cambridge University Press.

Aitchison, C., Macleod, N.E., and Shaw, S.J., 2000. *Leisure and tourism landscapes: Social and cultural geographies*. London: Routledge.

Altheide, D., Coyle, M., DeVriese, K., and Schneider, C., 2008. Emergent qualitative document analysis. In: S.N. Hesse-Biber and P. Leavy, eds. *Handbook of emergent methods.* New York: The Guilford Press. pp. 127-51.

Althusser, L., 1971. *On ideology*. Translated by B. Brewster, 2001. London and New York: Verso

Althusser, L., 1984. *On ideology*. New York: Verso.

Anderson, B., 2006. *Imagined communities*. New York: Verso.

Anderson, B., and Harrison, P., eds., 2010. *Taking-place: Non-representational theories and geography*. Surrey, England: Ashgate Publishing.

Anderson, B.R.O'G., 1983. *Imagined communities: Reflections on the origin and spread of nationalism*. London: Verso.

Armstrong, R.P., 1981. *The powers of presence: Consciousness, myth, and affecting presence.* Philadelphia: University of Pennsylvania Press.

Ashenburg, K., 2001. Traveling smart; On beaten paths or off, there's a guide to get you there. *New York Times.* [Online] 15 May. Available at: http://www.nytimes.com/2001/05/15/travel/traveling–smart–on–beaten–paths–or–off–there–s–a–guide–to–get–you–there.html?scp=1&sq=Traveling+Smart&st=nyt [Accessed: 28 September 2010].

Ashworth, G.J. and Tunbridge, J.E., 2000. *The tourist–historic city: Retrospect and prospect of managing a heritage city*. Amsterdam: Pergamon.

Ateljevic, I. and Doorne S., 2002. Representing New Zealand: Tourism imagery and ideology. *Annals of Tourism Research,* 29(3), pp. 648-67.

Atkinson, C.Z., 2004. Whose New Orleans? Music's place in the packaging of New Orleans for tourism. In: S.B. Gmelch, ed. *Tourists and tourism: A reader.* Long Grove, IL: Waveland Press, Inc. pp. 171-183.

Auge, M., 1995. *Non-places: Introduction to an anthropology of supermodernity*. New York: Verso.

Austin, J.L., 1975. *How to do things with words.* London: Oxford University Press.

Autostraddle., 2013. *Out of rage: Ani's not-so-righteous retreat.* Available at: <http://www.autostraddle.com/out-of-rage-anis-not-so-righteous-retreat-214869/>.

Baerenholdt, J.O. and Granas, B. eds., 2008. *Mobility and place: Enacting northern European peripheries*. Aldershot, UK: Ashgate Publishing.

Baerenholdt, J.O., Haldrup, M., Larsen, J., and Urry, J., 2004. *Performing tourist places*. Burlington, VT: Ashgate Publishing.

Bar, D. and Cohen-Hattab, K., 2003. A new kind of pilgrimage: The modern tourist pilgrim of nineteenth-century and early twentieth century Palestine. *Middle East Studies,* 39(2), pp. 131-148.

Barker, C. and Galasinski D., 2001. *Cultural studies and discourse analysis: A dialogue on language and identity.* Thousand Oaks, CA: Sage.

Barthes, R., 1972. *Mythologies.* Translated by A. Lavers. New York: Hill and Wang.

Bateson, G., 1972. A theory of play and fantasy. In: 2000. *Steps to an ecology of mind.* Chicago, IL: University of Chicago Press. pp. 177-95.

Baudrillard, J., 1994. *Simulacra and simulation.* Ann Arbor, MI: University of Michigan Press.

Bauman, Z., 1996. From pilgrim to tourist: Or a short history of identity. In: S. Hall and P. du Gay, eds. *Questions of cultural identity.* London: SAGE. pp. 18-36.

Beeton, S., 2004. Rural tourism in Australia—has the gaze altered? Tracking rural images through film and tourism promotion. *International Journal of Tourism Research,* 6(3), pp. 125-35.

Beiner, R., 1983. *Political judgment.* Chicago, IL: University of Chicago Press.

Benson, M., 2011. The movement beyond (lifestyle) migration: Mobile practices and the constitution of a better way of life. *Mobilities,* 6(2), pp. 221-35.

Benson, M. and O'Reilly, K. 2009. Migration and the search for a better way of life: A critical exploration of lifestyle migration. *The Sociological Review,* 57(4), pp. 608-25.

Birthright Israel., 2008. *Birthright Israel.* Available at: http://www.birthrightisrael.com/site/PageServer?pagename=about_faq#23.

Bitzer, L.F., 1959. Aristotle's enthymeme revisited. *Quarterly Journal of Speech,* 45, pp. 399-408.

Bitzer, L.F., 1968. The rhetorical situation. *Philosophy and Rhetoric,* 1, pp. 1-14.

Black, E., 1965. *Rhetorical criticism: A study in method.* Madison, WI: University of Wisconsin Press.

Blunt, A. and Varley, A., 2004. Introduction: Geographies of home. *Cultural Geographies,* 11(1), pp. 3-6.

Boorstin, D.J., 1992 [1961]. *The image: A guide to pseudo-events in America.* New York: Vintage.

Bowman, M.S., and Pezzullo, P.C., 2009. What's so "dark" about "dark tourism"?: Death, tours, and performance. *Tourist Studies*, 9(3), pp. 187-202.

Bremer, T.S., 2004. *Blessed with tourists: The borderlands of religion and tourism in San Antonio.* Chapel Hill: North Carolina Press.

Briggs, C.L., 1986. *Learning how to ask: A sociolinguistic appraisal of the role of the interview in social science research.* Cambridge, UK: Cambridge UP.

Brin, E., 2006. Politically-oriented tourism in Jerusalem. *Tourist Studies*, 6(3), pp. 215-43.
Brin, E. and Noy, C., 2010. The said and the unsaid: Performative guiding in a Jerusalem neighbourhood. *Tourist Studies*, 10(1), pp. 19-33.
Brown, G. and Yule, G., 1983. *Discourse analysis*. Cambridge, UK: Cambridge UP.
Bruner, E.M., 2001. The Massai and the Lion King: Authenticity, nationalism and globalization in African tourism. *American Ethnologist,* 28(4), pp. 881-908.
Bruner, E.M., 2005. *Culture on tour: Ethnographies of travel.* Chicago, IL: University of Chicago Press.
Buchmann, A., Moore, K. and Fisher, D., 2010. Experiencing film tourism: Authenticity and fellowship. *Annals of Tourism Research,* 37(1), pp. 229-48.
Carlson, A., 1985. *Dialogue games: An approach to discourse analysis*. Dordrecht, The Netherlands: D. Reidel Publishing.
Cartier, C. and Lew, A., eds., 2005. *Seductions of place: Geographical perspectives on globalization and touristed landscapes.* London: Routledge.
Casey, E.S., 1993. *Getting back into place: Toward a new understanding of the place-world.* Bloomington, IN: Indiana University Press.
Casey, E.S., 1997. *The fate of place: A philosophical history.* Berkeley, CA: University of California Press.
Chambers, E., 2000. *Native tours: The anthropology of travel and tourism.* Long Grove, IL: Waveland Press, Inc.
Chaney, D., 1993. *Fictions of collective life.* London: Routledge.
Chard, C. and Langdon, H., eds., 1996 *Transports: Travel, pleasure and imaginative geography, 1600-1830*. New Haven, CT: Yale University Press.
Chidester, D. and Linenthal, E.T., eds., 1995. *American sacred space*. Bloomington, IN: Indiana University Press.
Childs, G., 1999. The bird. In: *30 years of "Climbing" magazine*. Carbondale, CO: Primedia Special Interest Publications. pp. 241-43.
Chronis, A., 2005. Constructing heritage at the Gettysburg storyscape. *Annals of Tourism Research,* 32(2), pp. 386-406.
Chronis, A., 2008. Co-constructing the narrative experience: Staging and consuming the American Civil War at Gettysburg. *Journal of Marketing Management,* 24(1/2), pp. 2-27.
Chronis, A., 2012. Between place and story: Gettysburg as tourism imagery. *Annals of Tourism Research,* 39(4), pp. 1797-1816.
Clark, N., 2003. The play of the world. In: M. Pryke, G. Rose and S. Whatmore, eds. *Using social theory: thinking through research.* London: SAGE. pp. 28-46.
Clifford, J., 1997. *Routes: Travel and translation in the late twentieth century.* Cambridge, MA: Harvard University Press.
Cloke, P. and Perkins, H.C., 1998. "Cracking the canyon with the awesome foursome": Representations of adventure in New Zealand. *Environment and Planning D: Society and Space,* 16(2), pp. 185-218.

Cloke, P. and Perkins, H.C., 2002. Commodification and adventure in New Zealand tourism. *Current Issues in Tourism,* 5(6), pp. 521-49.

Cohen, A.P., 1985. *The symbolic construction of community.* London: Elias Horwood.

Cohen, E., 1973. Nomads from affluence: Notes on the phenomenon of drifter-tourism. *International Journal of Comparative Sociology,* 14(1), pp. 89-104.

Cohen, E., 1979. A phenomenology of tourist experience. In: S. Williams, ed. *Tourism: Critical concepts in the social sciences.* London: Routledge. Vol. 2 pp. 3-26.

Cohen, E., 1992. Pilgrimage and tourism: Convergence and divergence In: A. Morinis, ed. *Sacred journeys: The anthropology of pilgrimage.* pp. 47-61.

Cohen, E., 2004. Backpacking: diversity and change. In: G. Richards and J. Wilson, eds. *The global nomad: Backpacker travel in theory and practice.* Clevedon, UK: Channel View. pp. 43-59.

Cohen, S.A., 2010a. Chasing a myth? Searching for "self" through lifestyle travel. *Tourist Studies,* 10(2), pp. 117-33.

Cohen, S.A., 2010b. Personal identity (de)formation among lifestyle travellers: A double-edged sword. *Leisure Studies,* 29(3), pp. 289-301.

Cohen, S.A., 2011. Lifestyle travelers: Backpacking as a way of life. *Annals of Tourism Research,* 38(4), pp. 1535-55.

Coleman, S. and Crang, M., eds., 2002. *Tourism: Between place and performance.* New York: Beghahn Books.

Condit, C.M. and Lucaites, J.L., 1993. *Crafting equality: America's Anglo-African word.* Chicago, IL: University of Chicago Press.

Conley, T.M., 1984. The enthymeme in perspective. *Quarterly Journal of Speech,* 70, pp. 168-87.

Connell, J., 2006. Medical tourism: Sea, sun, sand…and surgery. *Tourism Management,* 27(6), pp. 1093-1100.

Copenhagen Guidebook, 2009. *Copenhagen guidebook.* Copenhagen: Tourist Forlag.

Copenhagen Post, Sept. 25, 2009. 14:42. Available at: www.cphpost.dk/news/national/88-national/47028-visitdenmark-director-resigns.html. [Accessed: 1 December 2009].

Cosgrove, D., 1984. *Social formation and symbolic landscape.* Totowa, NJ: Barnes and Noble Books.

Cosgrove, D., 1985. Perspective and the evolution of the landscape idea. *Transactions of the Institute of British Geographers, New Series,* 10(1), pp. 45-62.

Crang, M., 1997. Picturing practices: Research through the tourist gaze. *Progress in Human Geography,* 21(3), pp. 359-73.

Crang, M., 1999. Knowing, tourism and practices of vision. In: D. Crouch, ed. *Leisure/tourism geographies: Practices and geographical knowledge.* London: Routledge. pp. 238-56.

Cresswell, T., 2002. Introduction: Theorizing place. In: G. Verstraete and T. Cresswell, eds. *Mobilizing place, placing mobility*. Amsterdam: Rodopi. pp. 11-32.

Cresswell, T., 2003. Landscape and the obliteration of practice. In: K. Anderson, M. Domosh, S. Pile and N. Thrift, eds. *Handbook of cultural geography*. London: SAGE. pp. 269-81.

Cresswell, T., 2006. *On the move: Mobility in the modern Western world*. London: Routledge.

Cresswell, T., 2010. Towards a politics of mobility. *Environment and Planning D: Society and Space,* 28(1), pp. 17-31.

Crouch, D., 2002. Surrounded by place: Embodied encounters. In: S. Coleman and M. Crang, eds. *Tourism: Between place and performance*. New York: Berghahn Books. pp. 207-18.

Crouch, D. 2005. Flirting with space: Tourism geographies as sensuous/expressive practice. In: C. Cartier and Lew, A., eds., *Seductions of place: Geographical perspectives on globalization and touristed landscapes,* London: Routledge. pp. 23-35.

Crouch, D., 2010. Flirting with space: Thinking landscape rationally. *Cultural Geographies*, 17(1), pp. 5-18.

Culler, J., 1981. Semiotics of tourism. *American Journal of Semiotics*. 1(1-2), pp. 127-40.

Davis, J.S., 2005. Representing place: Deserted isles and the reproduction of Bikini Atoll. *Annals of the Association of American Geographers*, 93(3), pp. 607-25.

Davis, R.C. and Marvin, G.R., 2004. *Venice, the tourist maze: A cultural critique of the world's most touristed city*. Berkeley, CA: University of California Press.

De Certeau, M., 1984. *The practice of everyday life*. Berkeley, CA: University of California Press.

Desmond, J.C., 1999. *Staging tourism: Bodies on display from Waikiki to Sea World*. Chicago, IL: The University of Chicago Press.

Dewalt, K.M. and Dewalt B.R., 2002. *Participant observation: A guide for fieldworkers*. Walnut Creek, CA: AltaMira Press.

Dewsbury, J.D., Harrison, P., Rose, M. and Wylie, J., 2002. Enacting geographies. *Geoforum,* 33(4), pp. 437-40.

Di Giovine, M.A., 2011. Pilgrimage: Communitas and contestation, unity and difference—An introduction. *Tourism Review*, 59(3), pp. 247-68.

Dreyfus, H.L., 1993. *Being-in-the-world. A commentary on Heidegger's Being and Time, Division 1*. Cambridge, MA: MIT Press.

Duncan, J. and Duncan, N., 1988. (Re)reading the landscape. *Environment and Planning D: Society and Space,* 6(1), pp. 117-128.

Durkheim, E., 1912. *The elementary forms of religious life*. Oxford, UK: Oxford University Press.

Eagleton, T., 1991. *Ideology: An introduction*. London and New York: Verso.

Eagleton, T., 1996. *Literary theory.* Minneapolis, MN: The University of Minnesota Press.
Ebron, P., 1999. Tourists as pilgrims: Commercial fashioning of transatlantic politics. *American Ethnologist,* 26(4), pp. 910-32.
Eco, U., 1976. Peirce's notion of interpretant. *MLN*, 91(6), pp. 1457-1472.
Eco, U., 1986. *Travels in hyperreality.* San Diego, CA: Harper Brace Jovanovich.
Edensor, T., 2000. Staging tourism: Tourists as performers. *Annals of Tourism Research*, 27(2), pp. 322-44.
Edensor, T., 2001. Performing tourism, staging tourism: (Re)producing tourist space and practice. *Tourist Studies,* 1(1), pp. 59-81.
Eliade, M., 1971. *The myth of the eternal return.* Princeton, NJ: Princeton University Press.
Esposito, R., 2010. *Communitas: The origin and destiny of community*. Translated by T. Campbell., Stanford, CA: Stanford University Press.
Farrell, T.B., 1993. *Norms of rhetorical culture*. New Haven, CT: Yale UP.
Fehér, I. and Kóródi M., [n.d]. *A vidéki turismus fejlesytése*. Budapest Szaktudás Kiadó Ház Zrt.
Filho, S.C., 2010. Rafting guides: Leisure, work and lifestyle. *Annals of Leisure Research,* 13(1/2), pp. 282-97.
Finnegan, C.A., 2005. Recognizing Lincoln: Image vernaculars in nineteenth-century visual culture. *Rhetoric and Public Affairs,* 8(1), pp. 31-57.
Foote, K.E., 1997. *Shadowed ground: America's landscape of violence and tragedy.* Austin, TX: University of Texas Press.
Fortier, A.M., 1999. Re-membering places and the performance of belonging(s). *Theory Culture Society,* 16(2), pp. 41-64.
Foucault, M., 1994. *The order of things: An archaeology of the human sciences*. New York: Vintage Books.
Franklin, A., 2004. Tourism as an ordering: Towards a new ontology of tourism. *Tourist Studies,* 4(2), pp. 227-301.
Friedman, A., 2004. *The machinery of talk: Charles Peirce and the sign hypothesis*. Stanford, CA: Stanford University Press.
Frow, J., 1991. Tourism and the semiotics of nostalgia. *October*, 57, pp. 123-51.
Gadamer, H.G., 1983. Hermeneutics as practical philosophy. Translated by F.G. Lawrence. *Reason in the age of science*, pp. 88-138.
Gaffey, S., 2004. *Signifying place: The semiotic realization of place in Irish product marketing.* Aldershot, UK: Ashgate.
Garrod, B. and Fyall, A., 1998. Beyond the rhetoric of sustainable tourism? *Tourism Management*, 19(3), pp. 199-212.
Germann Molz, J., 2008. Global abode: Home and mobility in narratives of round-the-world travel. *Space and Culture,* 11(4), pp. 325-42.
Gershon, I., 2010. Media ideologies: An introduction. *Journal of Linguistic Anthropology*, 20(2), pp. 283-93.
Giddens, A., 1984. *The constitution of society: Outline of the theory of structuration*. Berkeley, CA: University of California Press.

Giddens, A. and Turner, J., 1988. *Social theory today.* Stanford, CA: Stanford University Press.

Goffman, E., 1974. *Frame analysis: An essay on the organization of experience.* Boston, MA: Northeastern.

Goffman, E., 1979. Footing. *Semiotica,* 25(1-2), pp. 1-29.

Goffman, E., 1981. *Forms of talk.* Philadelphia, PA: University of Pennsylvania Press.

Gottdiener, M., 2001. *The theming of America: Dreams, media fantasies and themed environments.* Cambridge, UK: Westview.

Graburn, N., 2001. Secular ritual: A general theory of tourism. In: V. Smith, ed. *Hosts and guests: Tourism issues of the 21st century.* Elmsford, NY: Cognizant Communications. pp. 42-50.

Graburn, N.H.H., 1983. The anthropology of tourism. *Annals of Tourism Research,* 10(1), pp. 9-33.

Graburn, N.H.H., 2010. Secular ritual: A general theory of tourism. In: S.B. Gmelch, ed. *Tourists and tourism: A reader.* 2nd ed. Long Grove, IL: Waveland Press. pp. 25-36.

Gramsci, A., 1971. *Selections from the prison notebooks.* In: Q. Hoare and G.N. Smith, eds. New York: International Publishers.

Groth, P. and Bressi, T. eds., 1997. *Understanding ordinary landscapes.* New Haven, CT: Yale University Press.

Grosz, E., 1994. *Volatile bodies: Toward a corporeal feminism.* Bloomington, IN: Indiana University Press.

Gustafson, P., 2001. Roots and routes: Exploring the relationship between place attachment and mobility. *Environment and Behavior,* 33(5), pp. 667-86.

Haldrup, M. and Larsen, J., 2003. The family gaze. *Tourist Studies,* 3(1), pp. 23-45.

Hall, D., 1998. Tourism development and sustainability issues in Central and Southeastern Europe. *Tourism Management,* 9(5), pp. 423-31.

Hall, D., 2002. Brand development, tourism and national identity: The re-imaging of former Yugoslavia. *Journal of Brand Management,* 9(4/5), pp. 323-34.

Hall, S., 1995. New cultures for old? In: D. Massey and P. Jess, eds. *A place in the world? Places, cultures and globalization.* Oxford, UK: Oxford University Press. pp. 175-214.

Hannam, K., 2009. The end of tourism? Nomadology and the mobilities paradigm. In: J. Tribe, ed. *Philosophical issues in tourism.* Bristol, UK: Channel View Publications. pp. 101-13.

Harrison, P., 2006. 'How shall I say it …?' Relating the non-relational. *Environment and Planning A,* 38(4), pp. 590-608.

Haskins, Ekaterina V., 2003. Put your stamp on history: The USPS commemorative program celebrate the century. *Quarterly Journal of Speech,* 89(1), pp. 1-18.

Hayllar, B. Griffin, T. and Edwards, D., eds., 2008. *City spaces, tourist places: Urban tourism precincts.* Amsterdam: Elsevier.

Heidegger, M., 1971. Building dwelling thinking. Translated by A. Hofstadter. In: *Poetry, language, thought.* New York: Harper Colophon Books. pp. 141-60

Hine, C., 2008. Internet research as emergent practice. In: S.N. Hesse-Biber and P. Leavy, eds. *Handbook of emergent methods*. New York: The Guilford Press. pp. 525-41.

Hoelscher, S., 1998. *Heritage on stage: The invention of ethnic place in America's Little Switzerland.* Madison, WI: The University of Wisconsin Press.

Hookway, C., 1992. *Peirce*. London: Routledge.

Hopkins, J., 1998. Commodifying the countryside: Marketing myths of reality. In: C. Butler, C.M. Hall and J. Jenkins, eds. *Tourism and recreation in rural areas*. Chichester, UK: John Wiley and Sons. pp. 119-56.

Houser, N., 1992. Introduction. In: N. Houser and C. Klosel, eds. *The essential Peirce vol. 1.* Bloomington, IN: Indiana University Press, pp. xix-xli.

Ingold, T., 1993. The temporality of the landscape. *World Archaeology,* 25(2), pp. 152-74.

Ingold, T., 2000. *The perception of the environment: Essays on livelihood, dwelling and skill*. London: Routledge.

Jack, G. and Phipps, A., 2003. On the uses of travel guides in the context of German tourism to Scotland. *Tourist Studies*, 3(3), pp. 281-300.

Jaworski, A. and Pritchard, A., 2005. *Discourse, communication and tourism*. Clevedon, UK: Channel View Publications.

Jenkins, O.H., 2003. Photography and travel brochures: The circle of representation. *Tourism Geographies*, 5(3), pp. 305-328.

Johannesson, G.T. and Baerenholdt, J.O., 2008. Enacting places through the connections of tourism. In: J.O. Baerenholdt and B. Granas, eds. *Mobility and place: Enacting Northern European peripheries*. Aldershot, UK: Ashgate Publishing. pp. 155-66.

Jokela, S. and Linkola, H., 2013. 'State idea' in the photographs of geography and tourism in Finland in the 1920s. *National Identities,* 15(3), pp. 257-75.

Kelner, S., 2010. *Tours that bind: Diaspora, pilgrimage, and Israeli birthright tourism*. New York: New York University Press.

Kingsbury, P., 2011. Sociospatial sublimation: The human resources of love in Sandals Resorts International, Jamaica. *Annals of the Association of American Geographers*, 101(3), pp. 650-69.

Kingsbury, P., Crooks, V.A., Snyder, J., Johnston, R. and Adams, K., 2012. Narratives of emotion and anxiety in medical tourism: Comparing patient and academic perspectives. *Public Health Ethics,* 5(1), pp. 38-46.

Kirschenblatt-Gimblett, B., 1998. *Destination culture: Tourism, museums, and heritage.* Berkeley, CA: University of California Press.

Knight, C., 2012. Critic blasts Sistine Chapel's 'drunk cattle.' *Los Angeles Times*, Nov 2, 2012. Available at: http://articles.latimes.com/2012/nov/02/entertainment/la-et-cm-sistine-knight-20121102

Knudsen, B.T. and Waade, A.M., 2010. *Re-investing authenticity: Tourism place and emotions.* Bristol, UK: Channel View Publications.

Knudsen, D.C. and Gray, A.M., 2010. *Semiotics and advertising: What we would tell VisitDenmark® if they asked.* Paper presented at the Travel and Tourism Research Association Annual International Meetings, San Antonio, Texas, 2010.

Knudsen, D.C. and Greer, C.E., 2008. Heritage tourism, heritage landscapes, and wilderness preservation: The case of National Park Thy. *Journal of Heritage Tourism,* 3(1), pp. 18-35.

Knudsen, D.C. and Greer, C.E., 2011. Tourism and nostalgia for the pastoral on the island of Fyn, Denmark. *Journal of Heritage Tourism*, 6(2), pp. 87-98.

Knudsen, D.C. and Kotlen, M., 2006. Systems of production, globalization, international competitiveness, and prospects for the developing nations. In: D. Conway and N.C. Heynen, eds. *Globalization's contradictions*. New York: Routledge. pp. 65-75.

Knudsen, D.C., Metro-Roland, M., Soper, A.K., and Greer, C.E., eds. 2008. *Landscape, tourism, and meaning*. Aldershot, UK: Ashgate Publishing.

Knudsen, D.C. and Rickly-Boyd, J.M., 2012. Tourism sites as semiotic signs: A critique. *Annals of Tourism Research*, 39(2), pp. 1252-54.

Knudsen, D.C., Rickly-Boyd, J.M. and Greer, C.E., 2014. Myth, national identity, and the contemporary tourism site: The case of Amalienborg and Frederiksstaden. *National Identities,* 16(1), pp. 53-70.

Konecnik, M. and Go F., 2008. Tourism destination brand identity: The case of Slovenia. *The Journal of Brand Management,* 15(3), pp. 177-89.

Koshar, R., 1998. 'What ought to be seen': Tourists' guidebooks and national identities in modern Germany and Europe. *Journal of Contemporary History,* 33(3), pp. 323-40.

Kvale, S., 2008. *Doing interviews*. Thousand Oaks, CA: SAGE.

Larsen, J., 2005. Families seen sightseeing: Performativity of tourist photography. *Space and Culture,* 8(4), pp. 416-434.

Lau, R.W.K., 2011. Tourist sites as semiotic signs: A critical commentary. *Annals of Tourism Research,* 38(2), pp. 1252-54.

Laurier, E. and Philo, C., 2006. Possible geographies: A passing encounter in a cafe. *Area,* 38(4), pp. 353-63.

Leavy, P., 2008. Performance-based emergent methods. In: S.N. Hesse-Biber and P. Leavy, eds. *Handbook of emergent methods*. New York: The Guilford Press. pp. 343-59.

Lefebvre, H., 1991. *The production of space.* Translated by D. Nicholson-Smith. Oxford, UK: Blackwell Publishing.

Levi, D.S., 1995. The case of the missing premise. *Informal Logic,* 17(1), pp. 67-88.

Levi-Strauss, C., 1996. The principle of reciprocity. In: E.K. Aafke, ed. *The gift: An interdisciplinary perspective*. Amsterdam: Amsterdam University Press.

Lew, A.A., 1991. Place representation in tourist guidebooks: An example from Singapore. *Singapore Journal of Tropical Geography,* 12(2), pp. 14-137.

Lewis, P., 1979. Axioms for reading the landscape. In: D.W. Meinig, ed. *The interpretation of ordinary landscapes: Geographical essays*. New York: Oxford University Press. pp. 1-12.

Light, D., 2000. Gazing on communism: Heritage tourism and post-communist identities in Germany, Hungary, and Romania. *Tourism Geographies,* 2(2), pp. 157-76.

Light, D., 2001. Facing the future: Tourism and identity-building in post-socialist Romania. *Political Geography,* 20(8), pp. 1053-74.

Light, D., 2012. *The Dracula dilemma: Tourism, identity and the state in Romania.* Farnham, Surrey, England: Ashgate Publishing.

Liszka, J.J., 1996. *A general introduction to the semeiotic of Charles Sanders Peirce*. Bloomington, IN: Indiana University Press.

Lorimer, H., 2005. Cultural geography: the busyness of being "more-than-representational". *Progress in Human Geography,* 29(1), pp. 83-94.

Lury, C., 1997. The objects of traveling. In: C. Rojek and J. Urry, eds. *Touring cultures: Transformations of travel and theory.* London: Routledge. pp. 74-95.

MacCannell, D., 1976. *The tourist: A new theory of the leisure class*. New York: Schocken Books.

MacCannell, D., 1989. *The tourist: A new theory of the leisure class*. New York: Schocken Books.

MacCannell, D., 1992. *Empty meeting grounds: The tourist papers*. London: Routledge.

MacCannell, D., 1999. *The tourist: A new theory of the leisure class*. Berkeley, CA: University of California Press.

MacCannell, D., 2011. *The ethics of sightseeing*. Berkeley, CA: University of California Press.

Macpherson, H.M., 2010. Non-representational approaches to body-landscape relations. *Geography Compass,* 4(1), pp. 1-13.

Malpas, J.E., 1999. *Place and experience: A philosophical topography*. Cambridge, UK: Cambridge University Press.

Massey, D., 1991. A global sense of place. *Marxism Today,* June, pp. 24-29.

Massey, D., 1994. *Space, place, and gender.* Minneapolis, MN: University of Minnesota Press.

Massey, D., 2005. *For space*. London: Sage Publications.

Mauss, M., 1925 [2000]. *The gift: The form and reason for exchange in archaic societies*. New York: W.W. Norton and Co.

McGee, M.C., 1980. The "ideograph": A link between rhetoric and ideology. *Quarterly Journal of Speech*, 66(1), pp. 1-16.

McGregor, A., 2000. Dynamic texts and tourist gaze: Death, bones and buffalo. *Annals of Tourism Research*, 27(1), pp. 27-50.

McHugh, K.E., 2009. Movement, memory, landscape: An excursion in non-representational thought. *GeoJournal,* 74(1), pp. 209-18.

Meinig, D.W. ed., 1979. *The interpretation of ordinary landscapes: Geographical essays*. New York: Oxford University Press.

Mellor, D., 2001. *American rock: Region, rock, and culture in American climbing.* Woodstock, VT: The Countryman Press.

Memorial and Museum Auschwitz-Birkenau. 2012. *Record number of visitors to the Auschwitz Museum in 2011.* Available at: <http://en.auschwitz.org/m/index.php?option=com_content&task=view&id=953&Itemid=8>.

Metro-Roland, M.M., 2009. Interpreting meaning: An application of Peircean semiotics to tourism. *Tourism Geographies,* 11(2), pp. 270-79.

Metro-Roland, M.M., 2011. *Tourists, signs in the city: The semiotics of culture in an urban landscape.* Aldershot, UK: Ashgate Publishing.

Midtfyns Kommune, 2008. *Faaborg-Midtfyn.* Faaborg: Midtfyns Kommune.

Minca, C. and Oakes, T., eds., 2006. *Travels in paradox: Remapping tourism.* Lanham, MD: Rowman and Littlefield Publishers, Inc.

Morgan, N.F. and Gardiner E., eds. and trans., 1986. *The marvels of Rome.* 2nd ed. New York: Italica Press.

Morgan, N., and Pritchard, A., 2005. On souvenirs and metonymy: Narratives of memory, metaphor and materiality. *Tourist Studies*, 5(1), pp. 29-53.

Muirhead, L.R. and Rossiter, S., 1965. *The Blue Guides: England.* London: Ernest Benn Limited.

Nelson. V., 2005. Representation and images of people, place, and nature in Grenada's tourism. *Geografiska Annaler B,* 87(2), pp. 131-43.

Neumann, M., 1988. Wandering through the museum: Experience and identity in a spectator culture. *Border/Lines,* Summer, pp. 19-27.

Nordfyns Kommune, 2008. *Nordfynsguiden.* Bogense: Nordfyns Kommune.

Nöth, W., 1990. *Handbook of semiotics.* Bloomington, IN: Indiana University Press.

Nowicka, M., 2007. Mobile locations: Construction of home in a group of mobile transnational professionals. *Global Networks,* 7(1), pp. 69-86.

Noy, C. and Cohen, E., 2005. Introduction: Backpacking as a rite of passage in Israel. In: C. Noy and E. Cohen, eds., 2005. *Israel backpackers and their society: A view from afar.* Albany, NY: State University of New York. pp. 1-43.

Obenour, W.L., 2004. Understanding the meaning of the "journey" to budget travellers. *The International Journal of Tourism Research,* 6(1), pp. 1-15.

Okolie, A.C., 2003. Identity: Now you don't see it; now you do. *Identity*, 3(1), pp. 1-7.

Olins, W., 2002. Branding the nation: The historical context. *Journal of Brand Management,* 9(4/5), pp. 241-8.

Ooi, C-S., 2002. Contrasting strategies: Tourism in Denmark and Singapore. *Annals of Tourism Research,* 29(3), pp. 689-706.

Ormonde, M.E., 2012. Debunking the myth of the "Anchor Baby": Why proposed legislation limiting birthright citizenship is not a means of controlling unauthorized immigration. *Roger Williams University Law Review,* 17(3), pp. 861-87.

Ousby, I., 1990. *The Englishman's England: Taste, travel and the rise of tourism.* Cambridge, UK: Cambridge University Press.

Palmer, C., 2005. An ethnography of Englishness: Experiencing identity through tourism. *Annals of Tourism Research*, 32(1), pp. 7-27.

Parsons, N., 2005. Your city, my city, their city: Reflections on Budapest guidebooks. *The Hungarian Quarterly*, 46(Winter), pp. 82-100.

Peirce, C.S., 1931-58. *Collected papers of Charles Sanders Peirce* (CP) 8 vols. C. Hartshorne and P. Weiss, eds. Cambridge, MA: Harvard University Press.

Peirce, C.S., 1992-1998. *The essential Peirce: Selected philosophical writings* (EP) 2 vols. N. Houser and C. Kloesel, eds. Bloomington, IN: Indiana University Press.

Peirce, C.S. and Lady V. Welby, 1977. *Semiotic and significs: The correspondence between Charles S. Peirce and Lady Victoria Welby.* C.S. Hardwick and J. Cook, eds. Bloomington, IN: Indiana University Press. [SS]

Pezzullo, P.C., 2003. Touring "Cancer Alley", Louisiana: Performances of community and memory for environmental justice. *Text and Performance Quarterly,* 23(3), pp. 226-52.

Pezzullo, P.C., 2007. *Toxic tourism: Rhetorics of pollution, travel, and environmental justice*. Tuscaloosa, AL: University of Alabama Press.

Philips, A., 2002. *Prague: A novel.* New York: Random House.

Pons, P.O., 2003. Being-on-holiday: Tourist dwelling, bodies and place. *Tourist Studies,* 3(1), pp. 47-66.

Porter, D. and Prince, D., 1997. *Frommer's Scandanavia.* 17th ed. New York: MacMillan.

Poria, Y., Butler, R. and Airey, D., 2004. Links between tourists, heritage, and reasons for visiting heritage sites. *Journal of Travel Research,* 43(3), pp. 19-28.

Poria, Y., Butler, R. and Airey, D., 2006. Tourist perceptions of heritage exhibits: A comparative study of Israel. *Journal of Heritage Tourism,* 1(1), pp. 57-72.

Pratt, C., 1965. The south face of Mt. Watkins. *American Alpine Journal,* pp. 339-46.

Pretes, M., 2003. Tourism and nationalism. *Annals of Tourism Research,* 30(1), pp. 125-42.

Pritchard, A. and Morgan, N.J., 2001. Culture, identity and tourism representation: Marketing Cymru or Wales? *Tourism Management,* 22(2), pp. 167-79.

Ralph, D. and Staeheli, L.A., 2011. Home and migration: Mobilities, belongings and identities. *Geography Compass*, 5(7), pp. 517-30.

Rapport, N., 1994. "Busted for hush": Common catchwords and individual identities in a Canadian city. In: A. Amit-Talai and H. Lustiger-Thaler, eds. *Urban lives: fragmentation and resistance.* Toronto: McClelland and Stewart.

Relph, E., 1976. *Place and placelessness*. London: Pion.

Relph, E., 1981. *Rational landscapes and humanistic geography*. London: Croom Helm.

Ribeiro, M. and Marques, C., 2002. Rural tourism and the development of less favoured areas: Between rhetoric and practice. *International Journal of Tourism Research*, 4(3), pp. 211-20.

Rickly-Boyd, J.M., 2012. Authenticity and aura: A Benjaminian approach to tourism. *Annals of Tourism Research*, 39(1), pp. 269-89.

Rickly-Boyd, J.M., 2013a. Alienation: Authenticity's forgotten cousin. *Annals of Tourism Research*, 40(1), pp. 412-15.

Rickly-Boyd, J.M., 2013b. Existential authenticity: Place matters. *Tourism Geographies,* 15(4), pp. 680-686.

Rickly-Boyd, J.M., 2014. "It's supposed to be 1863 but it's really not": Inside the representation and communication of history at a pioneer heritage site. *International Journal of Heritage Studies.* (In press, available online).

Rickly-Boyd, J.M. and Metro-Roland, M., 2010. Background to the fore: The prosaic in tourist places. *Annals of Tourism Research,* 37(4), pp. 1164-80.

Righteous Babe Records. 2014. From Ani. Available at: <http://www.righteousbabe.com/blogs/news/11257877-from-ani>.

Righteous Babe Records. 2013. Righteous retreat cancelled. Available at: <http://www.righteousbabe.com/blogs/news/11177617-righteous-retreat-cancelled>.

Riley, P.J., 1988. Road culture of international long-term budget travelers. *Annals of Tourism Research,* 15(2), pp. 212-28.

Rojek, C. and J. Urry, 1997. *Touring cultures: Transformations of travel and theory*. London and NewYork: Routledge.

Rose, M., 2002. Landscape and labyrinths. *Geoforum,* 33(4), pp. 455-67.

Rowntree, L., 1996. Cultural landscape concept in American human geography. In: C. Earle, et al. *Concepts in human geography.* Lanham, MD: Rowan and Littlefield. pp. 127-60.

Sahlins, M.D., 1996. On the sociology of primitive exchange. In: E.K. Aafke, ed. *The gift: An interdisciplinary perspective.* Amsterdam: Amsterdam University Press.

Salter, M. and Bousfield, J., 2002. *The rough guide to Poland.* London: Rough Guide.

Sandall, R., 2011. Dreams of communitas. *Society*, 48(6), pp. 483-8.

Sauer, C.O., 1925. The morphology of landscape. *University of California Publications in Geography.* Berkeley, CA: University of California Press. pp. 19-54.

Saussure, F., 1959. *Course in general linguistics*. New York: Philosophical Library.

Saussure, F.D., 1983. *Course in general linguistics.* London: Duckworth.

Schiavenza, M., 2013. Hazy skies in Hong Kong? Just pose with a fake skyline *The Atlantic*. August 23, 2013 http://www.theatlantic.com/china/archive/2013/08/hazy-skies-in-hong-kong-just-pose-with-a-fake-skyline/278997/ [Accessed: 23 August 2013].

Schein, R.H., 1997. The place of landscape: A conceptual framework for interpreting an American scene. *Annals of the Association of American Geographers,* 87(4), pp. 660-680.

Schultz, P., 2003. *1000 places to see before you die.* New York: Workmen.

Scott, J.C., 1998. *Seeing like a state: How certain schemes to improve the human condition have failed.* New Haven, CT: Yale University Press.

Selwyn, T., 1996. *The tourist image: Myth and myth making in tourism.* Chichester, UK: Wiley.

Senda-Cook, S., 2012. Rugged practices: Embodying authenticity in outdoor recreation. *Quarterly Journal of Speech,* 98(2), pp. 129-52.

Shaffer, M.S., 2001. *See America first: Tourism and national identity, 1880-1940.* Washington, DC: Smithsonian Institution Press.

Shaffer, T.S., 2004. Performing backpacking: Constructing "authenticity" every step of the way. *Text and Performance Quarterly,* 24(2), pp. 139-60.

Sharpe, E.K., 2005. Delivering communitas: Wilderness adventure and the making of community. *Journal of Leisure Research*, 37(3), pp. 255-80.

Sheller, M. and Urry, J., 2006. The new mobilities paradigm. *Environment and Planning A,* 38(2), pp. 207-26.

Short, T.L., 2004. The development of Peirce's theory of signs. In: C. Misak, ed. *The Cambridge companion to Peirce.* Cambridge, UK: Cambridge University Press. pp. 214-40.

Simonsen, K., 2008. Place as encounters: Practice, conjunction, and co-existence. In: J.O. Baerenholdt and B. Granas, eds. *Mobility and place: Enacting Northern European peripheries*. Aldershot, UK: Ashgate. pp. 13-26.

Silvio, T., 2010. Animation: The new performance? *Journal of Linguistic Anthropology*, 20(2), pp. 422-38.

Smith, A., 2005. Conceptualizing city image change: The "re-imaging" of Barcelona. *Tourism Geographies*, 7(4), pp. 398-423.

Smith, A.D., 1991. *National identity*. Reno, NV: University of Nevada Press.

Smith, J.Z., 1987. *To take place: Toward theory in ritual*. Chicago, IL: University of Chicago Press.

Stebbens, R.A., 1997. Lifestyle as a generic concept in ethnographic research. *Quality and Quantity,* 31(2), pp. 347-60.

Stewart, A., 2005. Choosing guidebooks. In: J. Lorie and A. Sohanpaul, eds. 2006. *The traveler's handbook: The insider's guide to world travel.* London: Globe Pequot. pp. 589-93.

Taylor, J.E., 2010. *Pilgrims of the vertical: Yosemite rock climbers and nature at risk*. Cambridge, MA: Harvard University Press.

Tedman, G., 1999. Ideology, the state and the aesthetic level of practice. *Rethinking Marxism,* 11(4), pp. 57-73.

The Telegraph, 2013. Japan's tsunami-hit Fukushima nuclear plant to become tourist attraction. Available at: http://www.telegraph.co.uk/news/worldnews/asia/japan/10251717/Japans-tsunami-hit-Fukushima-nuclear-plant-to-become-tourist-attraction.html.

Theilmann, J.M., 1987. Medieval pilgrims and the origins of tourism. *Journal of Popular Culture*, 20(4), pp. 93-102.

Thrift, N., 1996. *Spatial formations*. London: SAGE.

Thrift, N., 1999. Steps to an ecology of place. In: D. Massey, J. Allen and D. Sarre, eds. *Human Geography Today*. Cambridge, UK: Polity Press. pp. 295-321.

Thrift, N., 2000. Afterwords. *Environment and Planning D: Society and Space,* 18(2), pp. 213-55.

Thrift, N., 2008. I just don't know what got into me: Where is the subject? *Subjectivity,* 22(1), pp. 82-9.

Todorov, T., 1990. *Genres in discourse.* New York and Melbourne: Cambridge University Press.

Travlou, P., 2002. Go Athens: A journey to the centre of the city. In: S. Coleman and M. Crang, eds. *Tourism. Between place and performance.* New York: Berghahn Books. pp. 108-27.

Tuan, Y.F., 1974. Space and place: Humanistic perspective. In: J. Agnew, D.N. Livingstone and A. Rogers, eds. 1996. *Human geography: An essential anthology*. Malden, MA: Blackwell Publishers. pp. 444-57.

Tuan, Y.F., 1975. Place: An experiential perspective. *Geographical Review,* 65(2), pp. 151-65.

Tuan, Y.F., 1977. *Space and place: The perspective of experience*. Minneapolis, MN: University of Minnesota Press.

Turner, V., 1969. *The ritual process: Structure and anti-structure*. Chicago, IL: Aldine.

Turner, V., 1973. The center out there: Pilgrim's goal. *History of Religions*, 12(3), pp. 191-230.

Turner, V. and Turner E., 1978. *Image and pilgrimage in Christian culture: Anthropological perspectives.* New York: Columbia University Press.

Urry, J., 1990. *The tourist gaze.* London: SAGE.

Urry, J., 2002. *The tourist gaze.* 2nd ed. London: SAGE.

Urry, J. and Larsen, J., 2011. *The tourist gaze.* 3rd ed. London: SAGE.

Van Aalst, I. and Boogaarts, I., 2002. From museum to mass entertainment. *European Urban and Regional Studies*, 9(3), pp. 195-209.

Van Gennep, A., 1960. *The rites of passage.* Chicago, IL: University of Chicago Press.

Walker, J., 1994. The body of persuasion: A theory of the enthymeme. *College English,* 56(1), pp. 46-65.

Walton, D.N., 2001. Enthymemes, common knowledge, and plausible inference. *Philosophy and Rhetoric,* 34, pp. 93-112.

Wang, N., 1999. Rethinking authenticity in tourism experience. *Annals of Tourism Research,* 26(2), pp. 349-70.

Wheaton, B., 2004. Introduction: Mapping the lifestyle sport-scape. In: B. Wheaton, ed. *Understanding lifestyle sports: Consumption, identity and difference*. London: Routledge. pp. 1-28.

Wilson, C. and Groth, P., 2003. *Everyday America: Cultural landscape studies after J.B. Jackson*. Berkeley, CA: University of California Press.

Wilson, J., 2009. Tourism geographies: Space, place and lifestyle mobilities. *Tourism Geographies,* 11(1), pp. 124-6.

Winchester, H.P.M., 2005. Qualitative research and its place in human geography. In: I. Hay, ed. *Qualitative research methods in human geography.* 3rd ed., Oxford, UK: Oxford University Press. pp. 3-17.

Wylie, J., 2005. A single day's walking: Narrating self and landscape on the South West Coast Path. *Transactions of the Institute of British Geographers,* 30(2), pp. 234-47.

Wylie, J., 2007. *Landscape*. London: Routledge.

Xue, L., Manuel-Navarrette, D. and Buzinde, C.M., 2014. Theorizing the concept of alienation in tourism studies. *Annals of Tourism Research,* 44(1), pp. 186-99.

Young, I.M., 1990. *Justice and the politics of difference*. Princeton, NJ: Princeton University Press.

Young, M., 1999. The social construction of tourist places. *Australian Geographer*, 30(3), pp. 373-89.

Yuval-Davis, N., 2006. Belonging and the politics of belonging. *Patterns of Prejudice,* 40(3), pp. 197-214.

Yudice, G., 2003. *The expediency of culture: Uses of culture in the global era.* Durham, NC: Duke University Press.

Žižek, S., 1989. *The sublime object of ideology*. London and New York: Verso.

Index

Locators given in *italics* refer to figures.